The Social Challenges and Opportunities of Low-Carbon Development

This book explores the social implications and challenges of low-carbon development. The argument of the book is that a broad understanding of low-carbon development is essential for mitigating climate change and enabling development in a carbon-constrained world, but there are risks that low-carbon development might come at a price that is both social and economic. These risks need to be carefully assessed and reduced. The main aim of the book is to explore, critically analyse and propose different ways of understanding low-carbon development from a social perspective in both developed and developing countries. The author uses concepts such as low-carbon development, social policy, sustainable development and environmental justice to understand the social implications of low-carbon development projects.

The book first elaborates the need to understand the social issues and challenges of low-carbon development in both developed and developing countries. It then discusses five contemporary challenges of low-carbon development:

1 the social consequences of Chinese hydropower dams in the Mekong region;
2 the cost of the transition to renewable energies such as wind energy in Germany;
3 the challenges of carbon offsetting in Brazil;
4 the nexus of fuel-inefficient housing and fuel poverty in England;
5 solar power for refugees in Africa.

The book fills a crucial gap for researchers, postgraduates, practitioners and policy-makers in the fields of climate change, development and social policy.

Johan Nordensvärd is a Lecturer in Social Policy at the University of Southampton, UK.

Routledge Studies in Low Carbon Development

Low Carbon Transitions for Developing Countries
Frauke Urban

Towards Low Carbon Cities in China
Urban form and greenhouse gas emissions
Sun Sheng Han, Ray Green and Mark Wang

The Political Economy of Low Carbon Transformation
Breaking the habits of capitalism
Harold Wilhite

The Social Challenges and Opportunities of Low-Carbon Development
Johan Nordensvärd

The Social Challenges and Opportunities of Low-Carbon Development

Johan Nordensvärd

LONDON AND NEW YORK

First published 2017
by Routledge

2 Park Square, Milton Park, Abingdon, Oxfordshire OX14 4RN
52 Vanderbilt Avenue, New York, NY 10017

Routledge is an imprint of the Taylor & Francis Group, an informa business

First issued in paperback 2019

© 2017 Johan Nordensvärd

The right of Johan Nordensvärd to be identified as author of this work has been asserted by him in accordance with sections 77 and 78 of the Copyright, Designs and Patents Act 1988.

All rights reserved. No part of this book may be reprinted or reproduced or utilised in any form or by any electronic, mechanical, or other means, now known or hereafter invented, including photocopying and recording, or in any information storage or retrieval system, without permission in writing from the publishers.

Trademark notice: Product or corporate names may be trademarks or registered trademarks, and are used only for identification and explanation without intent to infringe.

British Library Cataloguing-in-Publication Data
A catalogue record for this book is available from the British Library

Library of Congress Cataloging-in-Publication Data
A catalog record for this book has been requested

ISBN: 978-0-415-73836-1 (hbk)
ISBN: 978-0-367-24796-6 (pbk)

Typeset in Sabon
by Wearset Ltd, Boldon, Tyne and Wear

Contents

Illustrations

Figure

Tables

Box

Acknowledgements

First of all, I would like to thank the patient editors at Routledge for being very helpful during the writing of the book. My thanks go especially to Helen Bell.

Second, I would like to thank my co-authors who have helped to make this book possible. I would particularly like to thank Frauke Urban, Tamra Gilbertson, Domiziana Marinelli and Carolyn Snell for their hard work.

Third, I would like to thank Larry Lohman and Jutta Kill for their insightful comments on Chapter 5, which have been very helpful.

I would also like to thank my family and especially my daughter for making my days brighter and more exciting.

Introduction

Johan Nordensvärd

The idea for this book started from a famous quote from Max Weber's seminal book *The Protestant Ethic and the Spirit of Capitalism*. One hundred and ten years later, the quote is just relevant today as it was 1905. He brought forward a central doubt that market capitalism can regulate itself and use less natural resources than needed. He wrote that the capitalist economic world order will proceed until "the last ton of fossilized coal is burnt" (Weber, 1953: 181). Scholars such as Walden Bello (2008) and Gus Speth (2008) argue that we are heading towards either a collapse of the present capitalist system or a collapse of our global climate. Nevertheless, the pessimism that Weber had towards Western modernism and capitalism has often been overshadowed by the number of people who share an optimistic view of the future, according to which things get better and better every year.

This optimism has been inherent in the way that modernism and the Enlightenment have understood development and progress. The meta-narrative of modernity had argued for the power of reason over ignorance, the power of order over disorder and the power of science over superstition. The old ruling classes with their archaically ordered society could be replaced with capitalism as a new mode of production and a transformation of the social order. This was the foundation upon which, it was argued, humanity would be able to achieve progress. It is a quest for the application for reason and for emancipation from ignorance, insecurity and violence. The main mechanism of this meta-narrative is that there is a larger movement towards human emancipation, despite setbacks and anomalies like wars, disasters and injustices; society is always moving forwards (Miller and Real, 2000: 19).

Climate change is a challenge to this optimistic view of development and capitalism. In the light of climate change, the discourse around development has started to shift focus away from pure growth towards a transformation. Despite the bleak outlook of climate change, and the loss of biodiversity and intact ecosystems, there is a new optimism that reason and progress can transform capitalism and thereby prevent that "last ton of fossilized coal" from being burnt.

Nevertheless, the concept of growth and consumption is ingrained in the way that we perceive our world: "Someone once said that it is easier to imagine the end of the world than to imagine the end of capitalism" (Jameson, 2003: 76). Capitalism and its commodification process have had not only environmental implications but also severe social implications for large numbers of the population. One of the more destructive aspects of capitalism was seen in the nineteenth century when some sectors of the population in the Western world became dependent upon the market for the survival. Karl Polanyi therefore discussed labour as a "fictitious commodity" since it is not produced for sale and it cannot be detached from the rest of a human's life (Polanyi, 1944: 72). Decommodification has often been linked to being a citizen in a welfare state and thereby linked to both duties and rights towards the state. The first conception of decommodification, as discussed by Polanyi, "protected citizens from major social risks and insulated their living standards from dependence on wage payments" and "the counter-movement that pressed for social reforms led to the creation of a welfare state dependent on public services paid for by taxes and social contributions" (Gough, 2010: 62). The welfare state moderated and mitigated the negative social implications of capitalism in the Western world.

Nevertheless, Polanyi did point out the importance of adding environment to the analysis. He argued that land is also a fictitious commodity. Land is considered by Polanyi as "another name for nature, which is not produced by man" (1944: 72). He argued that the commodification of land, natural resources, the oceans, and so forth will generate collective 'bads'. This will need a collective response from society. Polanyi argued for a more active role for the state in regulating the land and protecting natural resources from market forces: "[T]he commodity fiction disregarded the fact that leaving the fate of soil and people to the market would be tantamount to annihilating them" (Polanyi, 1944: 73). If labour and land are fictitious commodities, then how can we prevent them from being excessively exploited on a global scale? In many countries there is neither a strong welfare state to mitigate the exploitation of human labour nor an effective regime to protect the environment.

The question now is whether capitalism can prevent climate change and engage in poverty reduction at the same time. We should be under no illusion that human wealth has been created on the back of people and the environment so there are of course many question marks over how the exploitation of people and land can be prevented while still promoting growth. According to a narrow understanding of low-carbon development, there is a belief that both poverty reduction and climate change mitigation can be brought about with green capitalism and green economic growth. Progress will green everything eventually.

A narrow understanding of low-carbon development (or, perhaps more accurately, low-carbon growth) does imply at first sight a rather harmonious marriage between climate-friendly activities and international development.

One might of course think that a low-carbon economy and development are not mutually exclusive but that they could be combined in both bold and progressive ways. Low-carbon development is often seen as a possible answer to the question of how humanity is going to tackle its excessive carbon dioxide (CO_2) emissions without sacrificing either economic and social development, giving the best of both worlds. In general, low-carbon development is a development model that is based on climate-friendly low-carbon energy and follows principles of sustainable development, makes a contribution to the avoidance of dangerous climate change and adopts patterns of low-carbon consumption and production (Skea and Nishioka, 2008; Urban, 2010; Urban *et al.*, 2011).

The UK Department for International Development (DFID) has been seen as one of the original adopters of the concept of low-carbon development. The 2009 DFID White Paper *Eliminating World Poverty: Building Our Common Future* defines low-carbon development as "using less carbon for growth" by improving energy efficiency, using low-carbon energy sources, protecting natural resources that store carbon (such as forests and land), promoting low-carbon technologies and business models, and introducing policies and incentives that discourage carbon-intensive practices and behaviours (DFID, 2009: 58). But assuming we can decarbonise the economy, will this have a negative social impact? What are the social implications of low-carbon development projects? Is there a risk that equal access to environmental goods and equal exposure to environmental bads takes a back seat? This monograph will be a first attempt to look at these negative consequences from a conceptual perspective.

The argument of the book is that a broad understanding of low-carbon development is essential for mitigating climate change and enabling development in a carbon-constrained world, but that there are risks that low-carbon development might come at a price that is both social and economic. These risks need to be carefully assessed and reduced. The main aim of the book is to explore, critically analyse and propose different ways of understanding low-carbon development from a social perspective in both developed and developing countries. In low-income and lower-middle-income countries, issues of social justice and poverty reduction are the key to low-carbon development, while for higher-middle-income and high-income countries, low-carbon innovation and emission reductions are at the heart of implementing low-carbon development. Low-carbon development can bring opportunities and benefits for both developed and developing countries; nevertheless, low-carbon development must be understood through a social lens so that the poorest in every society will not suffer the costs of making our current world both sustainable and low carbon. The book will present both social implications and challenges.

There is currently no book that focuses completely on the social issues and challenges of low-carbon development. This book therefore aims to fill this gap by making a contribution to both teaching and research in terms of presenting the latest conceptual and empirical evidence on the social implications and issues of low-carbon development. The book will serve as a comprehensive introduction both to the social and economic aspects of development and to how these can be combined with low-carbon efforts. It aims to highlight the different challenges in both developed and developing countries. The book's approach is interdisciplinary and uses concepts such as low-carbon development, social policy, sustainable development and environmental justice to understand the social implications of low-carbon development projects. The book provides the reader with five relevant contemporary case studies that illustrate not just the issues and challenges but also the opportunities presented by low-carbon development.

The book first elaborates the need to understand the social issues and challenges of low-carbon development in both developed and developing countries. The book then discusses five contemporary challenges of low-carbon development: the social consequences of Chinese hydropower dams in the Mekong region; the cost of the transition to renewable energies such as wind energy in Germany; the challenges of carbon offsetting in Brazil; the nexus of fuel-inefficient housing and fuel poverty in England; and solar power for refugees in Africa.

References

Bello, W. (2008), "Will capitalism survive climate change?", *ZNet*, available at: http://focusweb.org/node/1197 (accessed 10 March 2016).

DFID (Department for International Development) (2009), *Eliminating World Poverty: Building our Common Future*, DFID White Paper, London: DFID.

Gough, I. (2010), "Economic crisis, climate change and the future of welfare states", *21st Century Society*, 5(1), 51–64.

Jameson, F. (2003), "Future city", *New Left Review*, 21 (May–June), 65–79.

Miller, G. and Real, M. (1998), "Postmodernity and popular culture: understanding our national pastime", in A. A. Berger (ed.), *The Postmodern Presence: Readings on Postmodernism in American Culture and Society*, Walnut Creek: AltaMira Press, pp. 17–34.

Polanyi, K. (1944), *The Great Transformation*, Boston: Beacon Press.

Skea, J. and Nishioka, S. (2008), "Policies and practices for a low-carbon society", *Climate Policy: Supplement – Modelling Long-Term Scenarios for Low-Carbon Societies*, 8(1), 5–16.

Speth, J. G. (2008), *The Bridge at the Edge of the World: Capitalism, the Environment and Crossing from Crisis to Sustainability*, New Haven, CT and London: Yale University Press.

Urban, F. (2010), "Pro-poor low carbon development and the role of growth", *International Journal of Green Economics*, 4(1), 82–93.

Urban, F., Mitchell, T. and Silva Villanueva, P. (2011), "Issues at the interface of disaster risk management and low-carbon development", *Climate and Development*, 3(3), 259–79.

Weber, M. (1953), *The Protestant Ethic and the Spirit of Capitalism*, New York: Scribner.

1 Low-carbon development

Its social implications and challenges

Johan Nordensvärd

Background

Today it is becoming more and more evident that the world is facing enormous challenges in terms of global environmental degradation and climate change. These two challenges are at the heart of concerns about international development. The recent interest in low-carbon development can be understood in the light of climate change and the ongoing global debate on how to mitigate the human impact on the climate on a global scale. Global climate change is not a distant vision of a troubled future, but very much a reality of today that requires urgent action. Former UN Secretary-General and President of the Global Humanitarian Forum Kofi Annan mentioned a few years ago that "Today, millions of people are already suffering because of climate change" (Annan, 2009: i).

The current UN Secretary-General Ban Ki-moon confirmed on a recent trip to the small Pacific nation of Kiribati that "climate change is not about tomorrow. It is lapping at our feet – quite literally in Kiribati and elsewhere" (Ban, 2011: 1). Ban continued: "I have watched the high tide impacting those villages. The high tide shows it is high time to act" (Ban, 2001: 1). He also addressed the current development model and suggested that something is "seriously wrong with our current model of economic development" (Ban, 2011: 1). The climate change discourse has been driven on the one hand by the scientific assessment reports of the Intergovernmental Panel on Climate Change (IPCC) and on the other hand by the climate policy process of the United Nations Framework Convention on Climate Change (UNFCCC).

While the scientific community has been working on researching and spreading knowledge around climate change for decades and the First Assessment Report of the IPCC dates back to 1990 (IPCC, 1990), the first highly influential report on climate change mitigation was not published until 2001 by the IPCC (IPCC, 2001). The publication, the IPCC's Third Assessment Report, included the so-called 'hockey stick diagram' by Michael Mann and colleagues (1999), which showed how both carbon emissions and average temperatures had increased significantly throughout the twentieth century, with the 1990s being the warmest decade of the millennium (IPCC, 2001). The diagram resembles the form of a hockey stick, hence its name. In the US, a controversy developed about the statistical methods underlying the research, fuelling debates between climate sceptics and non-climate sceptics. This was followed by more than a dozen scientific papers that confirmed the conclusions drawn by Mann and colleagues and the IPCC that the warmest decade in a millennium had most likely been at the end of the twentieth century. The urgent need to mitigate the emissions that lead to climate change was then acknowledged at a global level by the public.

The conclusions of IPCC research have been that climate change poses risks to humans, the environment and the economy (IPCC, 2013). The effects of climate change are reported to include rising temperatures, melting glaciers, sea-level rise, changes in precipitation, increases in extreme weather events like floods, droughts and cyclones, and acidification of the oceans (IPCC, 2007, 2013; Urban and Nordensvärd, 2013). Nevertheless, the impacts of climate change vary across different regions, intensities and scales. A degree of uncertainty is associated with climate change; however, there is consensus among the overwhelming majority of scientists about the anthropogenic (human-induced) causes of climate change, the main climatic impacts and their severity. It is well documented that so-called greenhouse gas emissions contribute to anthropogenic climate change (IPCC, 2007). There is a direct correlation between the increase of emissions of greenhouse gases, mainly CO_2, that lead to climate change and the rise of industrialisation, increasing affluence and consumption in developed countries (Urban and Nordensvärd, 2013; IPCC, 2013).

Greenhouse gases include CO_2, methane (CH_4), nitrous oxide (N_2O), hydrofluorocarbons (HFCs), perfluorocarbons (PFCs) and sulphur hexafluoride (SF_6) (UNFCCC, 1997). The most important greenhouse gas is CO_2, which is often only referred to as 'carbon', such as in relation to 'carbon emissions' and 'low-carbon development'. These greenhouse gases are emitted from the combustion of fossil fuels, from land-use changes and deforestation, from industrial activity and from transport (IPCC, 2007).

Scientists agree that the possibility of staying below the 2°C threshold between 'acceptable' and 'dangerous' climate change until the year 2100 is becoming less likely the longer that no serious global action on climate change is taken (Tyndall Centre, 2009; Richardson *et al.*, 2009; Urban, 2009; Urban *et al.*, 2011). A rise above 2°C before 2100 is likely to lead to abrupt and irreversible changes (IPCC, 2007). These changes could cause severe societal, economic and environmental disruptions which could severely threaten international development throughout the twenty-first century and beyond (Richardson *et al.*, 2009; Urban, 2010; Urban *et al.*, 2011; Urban and Nordensvärd, 2013). The IPCC's Fifth Assessment Report, published in 2014, confirmed that climate change is outpacing many earlier predictions (IPCC, 2014).

Just as important as the scientific discussion around climate change is the political discussion around climate change policy.

Climate change policy

The UNFCCC has become the global driver for climate change policy since its inception in the early 1990s. The UNFCCC developed as an international climate treaty at the Earth Summit in Rio de Janeiro in 1992 and thereby acknowledged the role of humans in contributing to climate change. The treaty aimed to prevent dangerous climate change, but made no commitment to emission reductions yet (UNFCCC, 1992).

This was followed by the Kyoto Protocol in 1997, which aimed to reduce greenhouse gas emissions to avoid dangerous climate change and had binding emission reduction commitments for developed countries for the first and second commitment periods of the Kyoto Protocol, 2008–12 and 2013–20 (UNFCCC, 1997). Due to Article 10 of the Protocol and the recognition of "common but differentiated responsibilities" of developed and developing countries for climate change, no emission reduction commitments were imposed on developing countries (UNFCCC, 1997: 9).

Climate change mitigation also had a prominent role in the Bali Action Plan and the Bali Roadmap of 2007. Mitigation was considered one of the five pillars of the Bali Action Plan (UNFCCC, 2007). In subsequent years, climate change mitigation has taken a prominent position in UN climate change negotiations and targets to mitigate emissions are one of the key elements over which major differences exist between developed and developing countries, but also within the group of developed and developing countries.[1]

This was apparent for the Copenhagen Accord in 2009, when no binding agreement could be reached on emission reduction and only the "strong political will to combat climate change" was mentioned (UNFCCC, 2009: 5). It was also apparent in the Cancun Agreements in 2010, when major advances were made in areas such as climate finance, technology transfer and REDD+ but only a pledge to "reduc[e] global greenhouse gas emissions so as to hold the increase in global average temperature below 2°C above preindustrial levels" could be agreed (UNFCCC, 2010: 3). The climate change conference in Durban in late 2011 delivered further evidence of just how hard it is to establish binding emission reduction targets for mitigating global climate change, as it was agreed to postpone any legally binding *global* agreement – beyond those obliged by the Kyoto Protocol – until at least 2015, with implementation by 2020 (UNFCCC, 2011). At Rio+20 in 2012, 20 years after the first Rio Earth Summit, climate change was still a hot but unresolved global issue. The 2015 Paris Agreement from COP21 showed that the global community is committed to tackling climate change. However, limiting global warming to 1.5–2°C will need serious international efforts.

There is an underlying conflict between developed countries which have historically been responsible for the large majority of carbon emissions, rising middle-income countries such as China, India, South Africa and Brazil with rising emissions, and low- and middle-income countries with low-carbon emissions both in the past and the present.

It is no secret that the consumption of wealthy developed countries is posing an environmental problem. At the Rio Summit in 1992, consumption patterns in the wealthy developed world were identified as a problem in the move towards a more sustainable society (Michaelis, 2000). A few years later, the World Wildlife Fund for Nature (WWF) argued that people "put pressure on forest, freshwater and marine ecosystems through the production and consumption of resources such as grain, fish, wood, and freshwater, and the emission of pollutants such as carbon dioxide" (WWF, 1998: 1). The Rio+20 Summit in 2012 highlighted the need to mitigate climate change and reduce the consumption of natural resources. Today, the main problem is the excessive consumption of energy and production of carbon emissions, particularly in developed countries, but also in emerging economies such as China, India, South Africa, Mexico and Brazil (Urban and Nordensvärd, 2013).

Low carbon and development

The main challenge is that there are wealthy countries that consume far too many resources, while the rest of the world dreams of modernising, which means consuming as much as developed countries. "In First World countries, industry's promise of unlimited consumerism has led to disproportionate levels of energy and water use, emission of greenhouse gases and the conversion of natural habitats" (Ho, 2006: 4). If the rest of the world were to aspire to Western levels of development and well-being, it would lead to drastic environmental consequences. There are doubts that global economic growth and social development can be sustained by the limited resources with which this planet can realistically supply us. Tim Jackson, for instance, suggests that "[t]here is no credible, socially just, ecologically sustainable scenario of continually growing incomes for a world of nine billion people" (Jackson, 2009: 57). Benjamin Barber expresses the ambivalence of developed industrial countries towards the demands of the developing countries to grow and develop: "this ecological consciousness has meant not only greater awareness but also greater inequality, as modernized nations try to slam the door behind them, saying to developing nations, 'The world cannot afford your modernization; ours has wrung it dry'" (Barber, 1992).

This situation becomes even more acute when we take into consideration the rise of climate change in the environmental discourse. Industrial development, as we know from developed countries, has often been both unsustainable and based around high carbon energy systems. "Development implies industrialization, urbanization and the intensification of resource use, the costs of which have often been externalized at the expense of the environment" (Ho, 2006: 4). It is therefore not a surprise that low-carbon development has become a popular approach to mitigate climate change but still promote economic development and to some degree preserve a belief that capitalist growth could be environmentally friendly. Low-carbon development does imply that the conflict between industrial development and climate change mitigation could be resolved through small reforms of the global capitalist system. By combining two rather disparate concepts, it implies that industrial development could eventually become low carbon and this could mean a decoupling of growth and carbon emissions.

Mainstream low-carbon development discourse tends to focus on pursuing development that is either low carbon or carbon neutral. "Low carbon development is situated at the interface of two major fields of study: climate change mitigation and international development" (Urban and Nordensvärd, 2013: 3). The concept of international development and development studies as a discipline is reported to have emerged in the late 1940s, 1950s and early 1960s. Development studies started out as a post-Second World War project of post-imperial intentional intervention in support of poorer 'developing countries'. 'Development' was driven by 'developed' Western/Northern countries and as such has often been accused of paternalism and trusteeship (Cowen and Shenton, 1996; Urban *et al.*, 2011).

Back in the 1950s, development policy was dominated by the universal goal of achieving modernity, by an optimistic worldview, by expecting the state to play an active and positive role and by focusing on national development (Humphrey, 2007; Urban *et al.*, 2011). There are various definitions for development; despite the universal use of the term, there is no universally agreed definition. Some scholars, such as Robert Chambers, define development simply as 'good change' (Chambers, 1995: 174); others make a distinction between formal development, such as development aid, and development as a deeper process of change, such as capitalism (Urban *et al.*, 2011). Gillian Hart makes a distinction between 'big D' Development and 'little d' development, whereby

> 'big D' Development (is) defined as a post-Second World War project of intervention in the 'Third World' that emerged in the context of decolonisation and the Cold War, and 'little d' development or the development of capitalism as a geographically uneven, profoundly contradictory set of historical processes.
>
> (Hart, 2001: 650)

There are various approaches to Western development thinking, such as rights-based approaches, which are focused on realising human rights and/or increasing the voice of marginalised groups or grassroots initiatives

(Mohan and Holland, 2001; Hickey and Mohan, 2005; Urban *et al.*, 2011); human development approaches, which incorporate broader development objectives than just economic ones and aim to expand human choices and strengthen human capabilities related to education, health, income and ensuring human rights (Jolly, 2003; Urban *et al.*, 2011); approaches that are based on concerns for the poorest 'bottom billion' (Collier, 2007; Urban *et al.*, 2011); and approaches that come from different disciplines such as anthropology, economics and political science and from different perspectives such as gender, globalisation and the environment. In other parts of the world, such as China, different streams of non-Western development thinking prevail which are more related to their own experience and philosophy of development (Urban *et al.*, 2011; Urban and Nordensvärd, 2013). Stuart Corbridge argues that development studies has an inherent conflict between the perceived areas of development in low- and middle-income countries and the implicit and sometimes explicit goal to reach the development level of high-income countries:

> Development studies is commonly understood to be committed both to a principle of difference (the Third World is different, hence the need for a separate field of studies) and a principle of similarity (it is the job of development policy to make 'them' more like 'us').
>
> (Corbridge, 2007: 179)

While development studies started with a great deal of optimism after the Second World War, the concept of development and development studies as a whole has had to endure criticism in recent years (Urban *et al.*, 2011). This is due to the continuation of widespread poverty in many parts of the world; states have become seen as part of the problem rather than part of the solution due to global neo-liberalism (Humphrey, 2007). Challenges like the global financial crisis, terrorism and large-scale environmental problems such as climate change are seen to require international and multilateral solutions (Urban *et al.*, 2011). One other major shift in development policy and practice relates to the 'Rising Powers': countries like China, India, Brazil, South Africa and states of the Middle East (Urban *et al.*, 2011). An aspect of this rise of new powers is a questioning of dominant 'Western' approaches to development (Humphrey, 2007; Urban and Nordensvärd, 2013).

The optimism of earlier decades has been replaced by a dose of pessimism, not least when development was declared dead in the 1990s by both the political right and the political left (Hart, 2001; Urban *et al.*, 2011). Fifteen years later, Gilbert Rist (2007) argued that development as perceived and practised by the West is 'toxic': "The essence of 'development' is the general transformation and destruction of the natural environment and of social relations in order to increase the production of commodities (goods and services) geared, by means of market exchange, to effective demand" (Rist, 2007: 488).

In many development institutes today, the notion of 'reimagining development' is prevalent as a need to rethink what development policy and practice means today, who is driving it and who will benefit. The Sustainable Development Goals (SDGs) have succeeded the Millennium Development Goals (MDGs) and indicate a merging of sustainability and development. The SDGs will set new targets for the time frame 2015–30 and will feature both climate and energy targets for international development.

A narrow definition of low-carbon development

Low-carbon development has become increasingly popular in recent decades but this does not mean that there is one unambiguous, agreed use of the concept. There is an inherent tension between a narrow concept (how we can make economic development and growth low carbon) and a broader understanding of low carbon and social transformation (how we can achieve social development within the constraints of our planet's finite resources). The classical, narrow definition of low-carbon development focuses only on reducing greenhouse gases in general and carbon emission in particular. This definition sees low-carbon development as a development model that is based on climate-friendly low-carbon energy and follows the principles of sustainable development, makes a contribution to the avoidance of dangerous climate change and adopts patterns of low-carbon consumption and production (Skea and Nishioka, 2008; Urban, 2010; Urban *et al.*, 2011).

Most of the content of low-carbon development has been around for longer than the actual concept. An example is ecological modernisation, which has become relatively widespread since it originated in Germany in the 1980s; it is also big in China (Christoff, 1996; Mol *et al.*, 2009). The principle is essentially about incorporating environmental interests, including climate change mitigation, within a successful capitalist economy. This implies 'policy integration' across environmental, economic and social decision-making (Nilsson and Eckerberg, 2007). Ecological modernisation entails a belief that through environmental changes to the economy it could be possible to "gradually decouple carbon emissions from economic activity" (Gough, 2011: 53).

While the academic community used other concepts and practitioners worked on renewable energy, technology transfer and innovation as separate issues and disciplines for decades, in the late 2000s the donor community – and particularly the UK Department for International Development (DFID) – first coined the term 'low-carbon development' to merge all of these disparate issues and disciplines together (DFID, 2009). The 2009 DFID White Paper *Eliminating World Poverty: Building Our Common Future* defines low-carbon development as "using less carbon for growth" by improving energy efficiency, using low-carbon energy sources, protecting natural resources that store carbon (such as forests and land), promoting low-carbon technologies and business models, and introducing policies and incentives that discourage carbon-intensive practices and behaviours (DFID, 2009: 58).

Low-carbon development was thus first developed by donors as a response to the immediate needs of developing countries to address climate change by mitigating emissions while ensuring development needs. The academic community has been relatively slow in responding to the emerging field of low-carbon development, although this has been changing in recent years. While there are a growing number of projects, activities, networks and publications by bilateral and multilateral donors on low-carbon development, there is a dearth of more academic and in-depth work in this field.

The dominant understanding of low-carbon development is narrow and is centred on how economic growth/economic development could become low carbon. Growth is often seen as the overriding goal of low-carbon development and growth relates back to the growth of the economy, the market and the overall wealth of people and societies. The narrow understanding of low-carbon development puts much faith in the capacity of technology to enable green growth. The major pillar in sustaining low-carbon growth or green growth relies very much upon technology being the lifeline of global capitalism. It is therefore no surprise that considerable numbers of books, articles and reports cover the development and spread of low-carbon technology. Many technological issues addressing low-carbon considerations have been hot topics since the 1970s and 1980s, such as technology transfer for low-carbon technologies and innovation for low-carbon energy, and their relevance and importance continue to this day. However, in the 1970s and 1980s these technologies were not called 'low carbon' but 'renewable energy' or 'clean energy' technologies. One significant focus has been, and is still today, on low-carbon energy transitions, which can be defined as shifts in a country's economic activities from an economy based on fossil fuels to one based (partially) on renewable and low-carbon energy. This means that switches take place from fossil fuel-based technologies to low-carbon technologies. Such transitions can take place in every sector of a country's or a region's economy (Urban, 2014).

The issue of mitigating greenhouse gas emissions only achieved momentum in the late 1990s and early 2000s, when the UN climate policy process began to fully develop. Many long-established fields addressing low-carbon considerations, such as hydropower and forestry, have seen a renaissance due to climate change and the emergence of low-carbon development. Technology is now seen as playing a central role in mitigating and adapting to climate change (Stern, 2006; IPCC, 2007) and for achieving low-carbon development. Innovation in low-carbon technology has become an important factor in policy discourse, but often innovation tends to occur only in developed countries.

It has therefore become increasingly important to support the transfer of low-carbon technologies to low- and middle-income countries. Both the UNFCCC and Kyoto Protocol have underwritten the need for rich developed nations to support and finance the transfer of low-carbon technologies to developing nations. Article 4.5 of the UNFCCC stated that developed countries had a duty to "take all practical steps to promote, facilitate and finance, as appropriate, the transfer of, or access to, environmentally sound technologies and know-how to other Parties, particularly developing country Parties, to enable them to implement provisions of the Convention" (UNFCCC, 1992, Article 4.5: 11). From a growth perspective, it has been seen as paramount that developing countries are able to achieve a "technological catch-up with the rest of the world" or "they will continue to fall behind other countries technologically and face deepening marginalisation in the global economy" (UNCTAD, 2007: i)

Low-carbon technology transfer is considered to be a means of promoting both growth and poverty reduction in a climate-friendly way in developing countries. Low-carbon technology is seen to both transform growth to become green and also to support poverty reduction. However, as with other low-carbon measures such as carbon markets, low-carbon technology transfers have not achieved their full potential and have been criticised. Schemes such as the Clean Development Mechanism aimed to encourage industrialised countries to reduce their emissions through the transfer of green technologies to developing countries (Grubb *et al.*, 2001; Ravindranath and Sathaye, 2002). Despite this, most of the cumulative investment (around 83 per cent) has reached rising middle-income countries such as China, India and Brazil instead of low-income countries (Byrne *et al.*, 2012). Seventy-five per cent of the investment went to five dominant low-emissions technologies: hydropower, methane avoidance, wind energy, biomass energy and landfill gas (Byrne *et al.*, 2012). Despite this, there is a predominant positive belief that by spreading low-carbon technology to developing countries, growth and development could happen without replicating the high carbon emission pathway of the developed countries today. This puts faith in the capacity of technological innovation to help developing countries to leapfrog over the polluting stages of historical industrial development.

Low-carbon development and its costs

The dominant view on low-carbon development relies not only upon technology to achieve green growth, but also upon adequate financing. Just as important is the discourse that nature needs to have a price attached and emissions/environmental degradations need to be represented as a direct cost. Influential studies such as The Economics of Ecosystems & Biodiversity (TEEB) publication *Mainstreaming the Economics of Nature* (TEEB, 2010) and the United Nations Environment Programme (UNEP) report and initiative *Towards a Green Economy* (UNEP, 2011) frame environmental problems in economic terms. The core argument in these studies is that the environment in general and carbon emissions in particular are to be understand in market terms and that a price needs to be attached to carbon emissions and environmental goods and services.

The idea that the environment is only valued by people, governments and firms if a price is attached to it became popular in the fields of ecological economics and environmental economics due to thinkers such as Robert Costanza and colleagues (1997). This thinking is based on the idea that nature provides environmental goods and services for free, such as clean air, clean water, clean soils and access to food, minerals and energy. As no price is attached to these goods and services, their value is often understated or non-existent. Environmental destruction and degradation, such as air pollution, water pollution, soil contamination, climate change, natural resource depletion and biodiversity loss have a 'zero' price tag and thereby lead to negative trade-offs and to a relationship of (self-)destructiveness that people have towards nature (TEEB, 2010). This understanding of not valuing nature has changed in recent years, since ecosystem services, nature conservation and even carbon emissions have become part of economic considerations (Urban and Nordensvärd, 2013).

The main problem is not capitalism or exploitation but rather that the systems have excluded environmental costs from the calculations. It will mean that the world will have to monetise the costs of carbon emissions on the one hand, but also the value of afforestation and other carbon sinks. Large elements of the dominant discourse around low-carbon development rely upon the costs of climate change. The benefits of early action on climate change are expected to outweigh the costs of such action, and certainly the costs of leaving climate change unchecked. The Stern Review estimates that unmitigated climate change could cost at least 5 per cent of global GDP per year, and up to 20 per cent of global GDP if more extreme climate predictions become reality (Stern, 2006). To stabilise global CO_2 equivalent emissions at 550 ppm[2], the Stern Review estimates that the annual costs for climate change mitigation could be as low as 1 per cent of global GDP by 2050. The Review considers 1 per cent of global GDP per year as significant costs; however, they are manageable (Stern, 2006). The Review estimates that each emitted tonne of CO_2 causes damage of about US$85, such as damage associated with sea-level rise, extreme weather events and rising temperatures. In comparison, the Review suggests that emissions can be cut at a cost of less than US$25 per tonne (Stern, 2006; Urban and Nordensvärd, 2013).

The Stern Review argues that a transition to a low-carbon economy is therefore urgently needed to stabilise greenhouse gas concentrations in the atmosphere. The Review argues further that a low-carbon economy poses challenges, but also offers opportunities for growth and competitiveness. Nevertheless, the Review's argument is that a transition to a low-carbon economy is only possible when effective policies and financial arrangements are in place, particularly for facilitating low-carbon transitions in developing countries. The Review suggests that a global low-carbon transition could eventually benefit the global economy by US$2.5 trillion per year. The Review further projects that by 2050, markets for low-carbon technologies could be worth at least US$500 billion (Stern, 2006). While mitigation is crucial, adaptation is also urgently needed to deal with unavoidable climatic impacts. In particular, adaptation efforts in developing countries need to be accelerated and supported, including through overseas development assistance (ODA) (Stern, 2006; Stern, 2007). A new report on climate economics has been published by the Global Commission on the Economy and Climate (2014) acknowledging the importance of mitigating climate change as early as possible.

More recent reports on the economics of climate change mitigation, such as those by the OECD (2009) and Ackerman and Stanton (2011), build on the findings of the Stern Review. Ackerman and Stanton (2011) present a report on the state of the art of climate economics in late 2011, which assesses the latest climate science, analyses the economic methodologies and assumptions used in climate economics and reiterates the warning messages on early action presented by the Stern Review. However, no new figures are presented. Similar assessments to the Stern Review have been undertaken on a regional basis for Southeast Asia (ADB, 2009) and Kenya (SEI, 2009). Managing environmental costs and possible trading with carbon emissions have been mainstreamed (Urban and Nordensvärd, 2013).

Does low-carbon development imply green growth?

In the dominant (and narrow) low-carbon development framework, growth is something that is not challenged by climate change and environmental degradation but very much compatible. There needs to be a considerable adjustment in how growth is perceived but "low carbon growth does not present a major challenge to traditional growth theory, it simply requires the internalisation of the environmental costs of growth through the appropriate pricing of goods and services" (Ellis *et al.*, 2009: 3). This approach takes for granted that the market could eventually be efficient in combating climate change (Clapp and Dauvergne, 2011). The DFID defines low-carbon development as using "less carbon for growth" (DFID, 2009: 58), which implies that economic growth is important and achievable, but needs to be low carbon.

In their DFID-funded report, Ellis and colleagues argue that "[c]onstraints on emissions raise the cost of energy which, in turn, reduces the output that can be achieved with a given set of inputs" (2009: 3). They conclude that the cost will be dependent upon the nature of policy and technological innovation. They hint towards technology being a saving grace for not just keeping growth at the present level:

> [p]olicies designed to promote green technological innovation and technology transfer could thus also potentially increase growth. In addition, some mitigation policies generate revenues (e.g. carbon taxes) and provide opportunities to stimulate growth through the judicious use of the revenues raised.
>
> (2009: 3)

Most of the discussion on low-carbon development still implies that both growth and capitalism are compatible with achieving low-carbon development. Low-carbon development addresses mitigation and development which requires an increased use of low-carbon energy and energy efficiency technology, protecting natural resources that store carbon such as forests, and implementing policies and incentives that discourage carbon-intensive practices and behaviours (DFID, 2009). The dominant perception is that growth and low-carbon development are not mutually exclusive. The Stern Review argues that it is not just about costs but that there could also be the promise of more growth in the future. The Review suggests that a global low-carbon transition could eventually benefit the global economy by US$2.5 trillion per year. The review further projects that by 2050, markets for low-carbon technologies could be worth at least US$500 billion (Stern, 2006).

Scholars such as Duncan Foley argue that it is about making the market work for the environment and adjusting the prices (Foley, 2006). William Nordhaus argues that there is a need to allocate a full set of property rights to the atmosphere, whereby carbon emissions get a price. This will force the external cost of carbon emissions onto the corporations and consumers (Nordhaus, 2007: 689). Servaas Storm argues: "The belief is that economic incentives, self-interest and market mechanisms will achieve an efficient, least-cost solution to the climate crisis" (Storm, 2009: 1017). The core argument in these studies is that the environment in general and carbon emissions in particular are to be understand in market terms and that a price needs to be attached to carbon emissions and environmental goods and services.

This understanding of not valuing nature has changed in recent years, since ecosystem services, nature conservation and even carbon emissions are part of economic considerations and today often have a price tag. This view is shared by organisations such as the International Energy Agency (IEA), the International Monetary Fund (IMF), the World Bank and the European Union (EU), all of which share a common belief in green growth (Storm, 2009: 1018). In a joint declaration on green growth, the OECD leaders highlight the importance of using the right policies:

> A number of well-targeted policy instruments can be used to encourage green investment in order to simultaneously contribute to economic recovery in the short term, and help to build the environmentally friendly infrastructure required for a green economy in the long term,

> noting that public investment should be consistent with a long-term framework for generating sustainable growth.
>
> (OECD, 2009: 1)

There are also alternative approaches to low-carbon development that focus on policy/politics and wider development issues, thereby going beyond a limited focus on technology and economics.

A broader understanding of low-carbon development

There is an argument that we need to see low-carbon development from a broader perspective than just technology, economic growth and industrial development. The narrow understanding of low-carbon development has become synonymous with green growth and technological innovation and far less about redesigning global, national and local societies and economies to be truly low carbon and socially just. We should of course not ignore a broader understanding of low-carbon development that aims to reshape societies to different degrees. Ecological modernisation, for example, implied a strong state that would drive the ecological development. Ecological modernisation had two major assumptions:

1 Low carbon technology will drive the development.
2 The state needs to invest to be able to boost its own competitiveness.

There is also a need "for integration across environmental, economic and social policy domains", which "will require a much more active state than the dominant model under neo-liberalism" (Gough, 2011: 53). This view is shared by green Keynesian scholars who merge social democratic and welfare state economics with low-carbon development and notions of a capitalist green economy.

Jeffrey Sachs argues that for a green economy "we will need large-scale public funding of research, development and demonstration projects; intellectual property rights to promote rapid dissemination to poor countries; and the promotion of public debate and acceptance of new options" (Sachs, 2008: 40). Joan Martinez-Alier argues that if "public investment must grow, as indeed it must to contain the rise in unemployment, it is better to channel it to the welfare of citizens and to 'green' energy production, than into motorways and airports" (Martinez-Alier, 2009: 1102). Hence, the state needs to fund education and research that will kick-start low-carbon development. Green Keynesian approaches could also go one step further and argue for a global carbon tax instead of carbon trading. This could be linked up with direct regulations and redistribution through a global welfare state (Bello and George, 2009).

Such an approach does not abolish the market, but regulates it and assumes that nation states at national and international level will coax the markets and firms to invest in a low-carbon economy and low-carbon development (Rezai *et al.*, 2009). There is a preference for financing this through global carbon taxes and/or Tobin taxes on international financial transactions. (A 'Tobin tax' refers to a tax on "short-term, cross-border foreign exchange transactions" (Wagner, 2004: 285).) It could also be financed through payments of developed industrial countries' carbon debts to developing countries (Bello and George, 2009).

A practical example of this could be the study of the Green Fiscal Commission in the UK that researched the effect of raising green taxes to 20 per cent of total tax revenues, which would be offset by lower employer social security contributions and an extra 10 per cent of funds for retrofitting houses and eco-innovation. The Commission suggests that this alone would achieve the UK's commitment to reduce greenhouse gases by 34 per cent by 2020. The study indicates further that a positive side effect could be a potential rise in employment due to lower employment costs (Green Fiscal Commission, 2009). A major issue is that high green taxes or high carbon taxes can be regressive since lower-income households tend to spend a higher share of their income on energy. Here, 'regressive' means that costs will be disproportionately high for people with lower incomes whereas people with higher incomes can more easily bear these costs. This is amplified by the fact that many people with lower incomes tend to live in fuel-inefficient houses, which cause higher energy costs. John Hills argues that carbon taxation requires complementary social policies on the one hand to invest in low-emission housing, transport and communities and on the other to protect low-income citizens who have high carbon consumption (Hills, 2009).

The steady-state economy has been one of the strongest policy frameworks for reducing humans' ecological footprints by creating limits to our consumption and the growth of the global economy. Herman Daly pointed out in the 1970s that an economy based around growth has also meant an "externalisation of environmental costs in growth accounting" and that there has been a "systematic underpricing of natural resources to the societal dominance of the other production factors such as capital and labour" (Altenburg and Pegels, 2012: 9). Daly argues that economists have focused too much on the economy's circulatory system and too little on its digestive tract. Growth of throughput will mean that more resources will be pushed through an overly narrow digestive tract. To continue the food metaphor, consumers should be transformed from gourmands to become gourmets, thereby focusing more on quality than on quantity (Daly, 2008: 1).

It is important to reduce growth and focus more on having an "economy with constant population and constant stock of capital, maintained by a low rate of throughput that is within the regenerative and assimilative capacities of the ecosystem" (Daly, 2008: 3). Daly argues that there are three key policy tools that need to be in place to transform societies towards such a steady-state economy:

1 minimum and maximum limits on income and wealth;
2 improvements to the tax system;
3 restrictions on population growth.

Daly has been one of the proponents of a maximum standard of living. In his words, a maximum standard of living should include both defined minimum and maximum limits on income and wealth. The key lies in creating new societal incentives when the maximum limit is reached. It is important that people are incentivised to "devote their further energies to noneconomic pursuits" so that confiscatory revenues would be rather small (Daly, 1977: 58) He argues that the lost opportunities for the wealthy should be made available for the less wealthy, who would pay taxes on their increased earnings. This would be less beneficial for the economic elite but it would be positive at "lower levels, leading to a broader participation in running the economy" (Daly, 1977: 56).

According to Daly, this is not just a national but a global scenario.

> The steady-state answer is that the rich should reduce their throughput growth to free up resources and ecological space for use by the poor, while focusing their domestic efforts on development, technical and social improvements that can be freely shared with poor countries.
>
> (Daly, 2008: 2–3)

An important dimension of maintaining both a minimum and maximum standard of living will be determined by social policy in general and its redistributive activities in particular. Daly and Joshua Farley highlight the importance of distribution of wealth and income to some degree both within and across countries and from an intergenerational perspective (Daly and Farley, 2009: 441). There is a difficult balance between on the one hand acknowledging what people have accumulated through their own efforts and on the other preventing people from being able to "capture for themselves values created by nature, by society, or by the work of others" (Daly and Farley, 2009: 441). It is also important that people pay "a fair price for what they receive from others, including the services provided by government, and for the costs they impose on others" (Daly and Farley, 2009: 441).

Daly's vision of the steady-state economy is very much dependent upon a reformation of the contemporary global and national tax systems in terms of what is being taxed, how it is taxed and what purpose a steady-state tax system should have. In theory, Daly wants to reform the system to be based on depletion quota instead of using policies such as pollution taxes (Daly, 1977). Daly and Farley advocate a highly progressive income tax that asymptotically approaches 100 per cent, more "direct limits on how much someone can earn, or relative limits that establish a legal ration between the highest and lowest income allowed" and a "high inheritance tax since much of the accumulated wealth is inherited" (Daly and Farley, 2009: 44). Daly discusses that the ideal would be to have an ecological tax reform, which means shifting the tax base away from just value added (income earned by labour and capital) towards the quality and nature of the value added and in the end how this contributes to the throughput flow. The focus lies in increasing the tax where the depletion of resources occurs (Daly, 2008: 8).

The discourse around 'degrowth' argues that endless economic growth is impossible to sustain. It questions the possibility of globally decoupling economic activity and emissions. In their famous book *The Limits to Growth* Donella Meadows and colleagues argue that exponential growth in population and material output threatens the well-being of all and that it could lead to an uncontrolled global decline (Meadows *et al.*, 1972). More recently, Tim Jackson (2009) and the Sustainable Development Commission (SDC) (2009) argue for "prosperity without growth" or prosperity within the ecological limits of our finite planet. Nevertheless, the concept of growth and consumption is ingrained in the way that we perceive our world. "Someone once said that it is easier to imagine the end of the world than to imagine the end of capitalism" (Jameson, 2003: 76). A degrowth approach would focus more on well-being and quality of life, which could be decoupled from growing Gross Domestic Product (GDP) or increased economic activity. This would imply a new macroeconomic approach that could mean, for example, a reduction in working hours (Victor, 2008).

However, the state would need to ensure that economic resources are more evenly distributed so that the poor would not be disproportionately affected. This might imply sharing some of the wealth of nations more equally, such as by taxing the richer groups of society more to ensure that the poorer groups of society are well off despite degrowth or reducing the excessive pay of top executives and creating employment opportunities with the available funding. Scholars such as Larry Lohman (2009), James Galbraith (2008) and Gus Speth (2008) argue not only for higher global carbon taxes and higher investment in low-carbon development, specifically low-carbon energy and low-carbon technology, but also suggest that there is a need to replace the market "by alternative democratic co-ordination and decision-making mechanisms" (Storm, 2009: 1026). One could therefore go one step further and suggest that a low-carbon society can only be sustained by low growth, no growth or, for Western developed countries, even degrowth. Scholars such as Walden Bello (2008) and Speth (2008) argue that we are heading towards either a collapse of the present capitalist system or a collapse of our global climate.

An alternative, though radical, scenario for global governance would be that natural resources are collectively owned and co-operatively managed and that the use of natural resources is subject to decentralised democratic decision-making. This would be dependent upon people and governments being willing to accept a model of lower consumption, lower growth, more distribution of resources and higher social equality that would result "in an improved welfare, a better quality of life and greater democratic control of production and (renewable) resources" (Storm, 2009: 1026). In the end, environmental socialism sees the environmental crisis as directly linked to the crisis of industrial capitalism. The environmental issues cannot be solved unless there is a radical restructuring of the international system (Clapp and Dauvergne, 2011). This would need a complete reconfiguration of our understanding of welfare states, global social policy and the global market.

Jim Skea and Shuzo Nishioka indicate that actions leading to low carbon development need to follow the principles of sustainable development and "ensur[e] that the development needs of all groups within society are met" (Skea and Nishioka, 2008: 6). Low-carbon development does not directly address issues of environmental sustainability in broader terms beyond climate change (Urban, 2011). There has been an attempt to broaden the scope of low-carbon development. Climate-compatible development is an example which aims for "development that minimises the harm caused by climate impacts, while maximising the many human development opportunities presented by a low emissions, more resilient, future" (Mitchell and Maxwell, 2011: 1). It essentially combines climate change adaptation, mitigation and development. Any low-carbon development can only be implemented when an adequate enabling environment is in place, which addresses the key political, economic, social and technological issues. One of the major challenges of low-carbon development is combining social justice and poverty reduction in developing countries while pushing for drastic emission reductions for developed countries.

This book should been seen to argue for a broader view of low-carbon development that includes a clear social dimension to the analysis. It is important that low-carbon development initiatives are not merely evaluated according to decreasing carbon emissions or increasing growth and development but also according to their social implications for people. It is important to analyse whether low-carbon development policies are socially sustainable and if they need to be coupled with social policy. Climate change, environmental degradation and environmental costs are bound to hit the poorest the hardest. This book argues that there is a need to set out a strong social framework to understand any attempts towards low-carbon development.

The argument

This monograph will not dwell too much on the harmonious relationship between low carbon and development or the economy. Rather, it aims to look at the social challenges that low-carbon development could imply. There is an ironic (and very social) twist in the fact that climate change hits the poorest the hardest but that the poorest are often those who are the least responsible for carbon emissions and contributing to climate change (Battisti and Naylor, 2009; Opschoor, 2008; Stern, 2009). The problem is that development comes at an environmental cost and environmental protection comes at a social cost. Meg Huby argues that

> [t]he problem is that, used alone, policies to promote social welfare carry long-term costs to both society and the environment, while policies for long-term environmental protection tend to produce distributive effects that work to the detriment of vulnerable groups of people in the shorter term.
>
> (Huby, 2001: 521–2)

On the other hand, one should also be careful not to assume that environmental and human exploitation are mutually exclusive; on the contrary, the exploitation of humans and the exploitation of the environment have gone very much hand in hand.

This book will be the first monograph on the social implications and challenges of low-carbon development. The book aims to elaborate the social implications and challenges of low-carbon development on the one hand and to address how low-carbon development can be achieved while also achieving poverty reduction and social justice on the other.

The argument of the book is that low-carbon development is essential for mitigating climate change and enabling development in a carbon-constrained world, but that there are risks that low-carbon development might come at a price that is both social and economic for the poorest in society. These risks need to be carefully assessed and reduced. The main aim of the book is to explore, critically analyse and propose different ways in which low-carbon development could become socially viable in both developed and developing countries.

The major challenge of low-carbon development is to bring opportunities and benefits for both developed and developing countries, as this would mean overcoming inherent issues of global social injustice and preventing the poorest in society from having to pay the bulk of the cost for both climate change and low-carbon development. The book will both present social implications and challenges as well as different strategies with which low-carbon development can be tackled in a socially just way at a global level. There is currently no book that focuses completely on the social issues and challenges of low-carbon development. This book therefore aims to fill this gap by making a contribution to both teaching and research in terms of presenting the latest conceptual and empirical evidence on the social implications and issues of low-carbon development. The book will serve as a comprehensive introduction both to the social and economic aspects of development and to how these can be combined with low-carbon efforts. It aims to highlight the different challenges in both developed and developing countries. The book uses in-depth case studies such as discussing what implications the energy transition in Germany has for energy prices and what the social implications of China's large hydropower dams are in China, as well as in Asia and Africa. The theoretical framework and the case studies will be presented in Chapter 2.

Notes

1 For example, there are major differences of opinion among developed countries, the so-called Annex I countries, particularly between the member states of the EU, which are mostly in favour of stringent emission reduction commitments, and countries like Canada, Japan, Russia and the US, which have either not ratified the Kyoto Protocol (US) or withdrawn from its second commitment period in 2011/2012 (Canada, Japan and Russia). There are also major differences of opinion among developing countries, the so-called non-Annex I countries, particularly between emerging emitters such as China and India – which would like to delay legally binding emission reduction targets for developing countries – and the Small Island Developing States and many African states – which would like to see stringent and immediate mitigation actions at a global level.

2 It should be noted that climate scientists estimate that for a 50 per cent chance of achieving the 2°C target to avoid dangerous climate change, a global atmospheric CO_2 equivalent concentration of 400 to 450 ppm needs to be achieved by 2100 (Richardson *et al.*, 2009; Pye *et al.*, 2010). Nevertheless, the Stern Review refers to the costs for stabilising CO_2 equivalent concentration at 550 ppm, which is more likely to result in a 3°C warming, rather than a 2°C warming.

References

Ackerman, F. and Stanton, E. A. (2011), *Climate Economics: The State of the Art*, Somerville: Stockholm Environmental Institute.

ADB (Asian Development Bank) (2009), *The Economics of Climate Change in Southeast Asia: A Regional Review*, Manila: ADB.

Altenburg, T. and Pegels, A. (2012), "Sustainability-oriented innovation systems: managing the green transformation", *Innovation and Development*, 2(1), 5–22.

Annan, K. (2009), *The Anatomy of a Silent Crisis: Human Impact Report – Climate Change*, Geneva: Global Humanitarian Forum.

Ban, K.-M. (2011), "Climate change lapping Pacific shores: Ban Ki-moon", *TVNZ*, available at: http://tvnz.co.nz/national-news/climate-change-lapping-pacific-shores-ban-ki-moon-4387638 (accessed 18 October 2012).

Barber, B. (1992), "Jihad Vs McWorld", *Atlantic Monthly*, March, available at: www.theatlantic.com/past/politics/foreign/barjiha.htm (accessed 30 April 2016).

Battisti, D. S. and Naylor, R. L. (2009), "Historical warnings of future food insecurity with unprecedented seasonal heat", *Science*, 323(5911), 240–4.

Bello, W. (2008), "Will capitalism survive climate change?", *ZNet*, available at: www.zmag.org/znet/viewArticle/17095 (accessed 10 March 2016).

Bello, W. and George, S. (2009), "A new, green, democratic deal", *The Transnational Institute*, available at: www.tni.org/en/article/a-new-green-democratic-deal (accessed 30 April 2016).

Byrne, R., Smith, A., Watson, J. and Ockwell, D. (2012), "Energy pathways in low-carbon development: the needs to go beyond technology transfer", in

D. Ockwell and A. Mallett (eds), *Low-Carbon Technology Transfer: From Rhetoric to Reality*, London: Routledge, pp. 123–42.

Chambers, R. (1995), "Poverty and livelihoods: whose reality counts?", *Environment and Urbanization*, 7(1), 173–204.

Christoff, P. (1996), "Ecological modernisation, ecological modernities", *Environmental Politics*, 5(3), 476–500.

Clapp, J. and Dauvergne, P. (2011), *Paths to a Green World: The Political Economy of the Global Environment*, second edition, Cambridge, MA: MIT Press.

Collier, P. (2007), *The Bottom Billion: Why the Poorest Countries are Failing and What Can Be Done About It*, Oxford: Oxford University Press.

Corbridge, S. (2007), "The (im)possibility of development studies", *Economy and Society*, 36(2), 179–211.

Costanza, R., d'Arge, R., de Groot, R., Farber, S., Grasso, M., Hannon, B., Limburg, K., Naeem, S., O'Neill, R., Paruelo, J., Raskins, R. G., Sutton, P. and van den Belt, M. (1997), "The value of the world's ecosystem services and natural capital", *Nature*, 387, 253–60.

Cowen, M. and Shenton, R. (1996), *Doctrines of Development*, London: Routledge.

DFID (Department for International Development) (2009), *Eliminating World Poverty: Building our Common Future*, DFID White Paper, London: DFID.

Daly, H. (1977), *Steady-State Economics: The Political Economy of Bio-Physical Equilibrium and Moral Growth*, San Francisco, CA: W.H. Freeman & Co.

Daly, H. (2008), *A Steady-State Economy: A Failed Growth Economy and a Steady-State Economy Are Not the Same Thing; They Are the Very Different Alternatives We Face*, London: SDC.

Daly, H. and Farley, J. (2009), *Ecological Economics: Principles and Applications*, Washington, DC: Island Press.

Ellis, K., Baker, B. and Lemma, A. (2009), *Policies for Low Carbon Growth: Executive Summary*, London: Overseas Development Institute (ODI).

Foley, D. (2006), *Adam's Fallacy: A Guide to Economic Theology*, Cambridge, MA: The Belknap Press.

Galbraith, J. K. (2008), *The Predator State: How Conservatives Abandoned the Free Market and Why Liberals Should Too*, New York: Free Press.

Global Commission on the Economy and Climate (2014), *Better Growth Better Climate: The New Climate Economy Report*, Washington: New Climate Economy, available at: http://newclimateeconomy.report/TheNewClimateEconomyReport.pdf (accessed 12 May 2016).

Gough, I. (2011), *Climate Change and Public Policy Futures*, New Paradigms in Public Policy project, London: British Academy.

Green Fiscal Commission (2009), *The Case for Green Fiscal Reform*, London: Green Fiscal Commission.

Grubb, M., Vrolijk, C. and Brack, D. (2001), *The Kyoto Protocol: A Guide and Assessment*, London: The Royal Institute of International Affairs.

Hart, G. (2001), "Development critiques in the 1990s: *cul de sac* and promising paths", *Progress in Human Geography*, 25(4), 649–58.

Hickey, S. and Mohan, G. (2005), "Relocating participation within a radical politics of development", *Development and Change*, 36(2), 237–62.

Hills, J. (2009), "Future pressures: intergenerational links, wealth, demography and sustainability", in J. Hills, T. Sefton and K. Stewart (eds), *Towards a More Equal Society? Poverty, Inequality and Policy since 1997*, Bristol: Policy Press, 319–39.

Ho, P. (2006), "Trajectories for greening in China: theory and practice", *Development and Change*, 37(1), 3–28.

Huby, M. (2001), "The sustainable use of resources on a global scale", *Social Policy and Administration*, 35(5), 521–37.

Humphrey, J. (2007), "Forty years of development research: transformations and reformations", *IDS Bulletin*, 38(2), 14–19.

IPCC (Intergovernmental Panel on Climate Change) (1990), *Climate Change: The IPCC Scientific Assessment*, First Assessment Report of the IPCC, Cambridge: Cambridge University Press, available at: www.ipcc.ch/ipccreports/far/wg_I/ipcc_far_wg_I_full_report.pdf (accessed 10 March 2016).

IPCC (Intergovernmental Panel on Climate Change) (2001), *Climate Change 2001: Working Group III – Mitigation*, Third Assessment Report of the IPCC, Cambridge: Cambridge University Press.

IPCC (Intergovernmental Panel on Climate Change) (2007), *Climate Change 2007: Synthesis Report*, Fourth Assessment Report of the Intergovernmental Panel on Climate Change, Cambridge: Cambridge University Press, available at: www.ipcc.ch/publications_and_data/publications_ipcc_fourth_assessment_report_synthesis_report.htm (accessed 30 April 2016).

IPCC (Intergovernmental Panel on Climate Change) (2013), *Working Group I Contribution to the Fifth Assessment Report: The Physical Science Basis*, Geneva: IPCC, available at: www.climatechange2013.org/images/uploads/WGIAR5_WGI-12Doc2b_FinalDraft_All.pdf (accessed 13 March 2016).

IPCC (Intergovernmental Panel on Climate Change) (2014), "Climate Change 2014: Mitigation of Climate Change", in IPCC, *Working Group III Contribution to the IPCC Fifth Assessment Report*. Cambridge: Cambridge University Press, available at: www.ipcc.ch/pdf/assessment-report/ar5/wg3/ipcc_wg3_ar5_full.pdf (accessed 8 July 2016).

Jackson, T. (2009), *Prosperity Without Growth*, London: Earthscan.

Jameson, F. (2003), "Future city", *New Left Review*, 21 (May–June), 65–79, available at: https://newleftreview.org/II/21/fredric-jameson-future-city (accessed 30 April 2016).

Jolly, R. (2003), "Human development and neo-liberalism: paradigms compared", in S. Kukudar-Parr and A. K. Shiva Kumar (eds), *Readings in Human Development*, New Delhi: Oxford University Press, pp. 82–92.

Lohmann, L. (2009), "Climate as investment", *Development and Change*, 40(6), 1063–83.

Mann, M. E., Bradley, R. S. and Hughes, M. K. (1999), "Northern hemisphere temperatures during the past millennium: inferences, uncertainties, and limitations", *Geophysical Research Letters*, 26(6), 759–62.

Martinez-Alier, J. (2009), "Socially sustainable economic de-growth", *Development and Change*, 40(6), 1099–119.

Meadows, D. H., Meadows, D. L., Randers, J. and Behrens III, W. W. (1972), *The Limits to Growth: A Report to the Club of Rome*, New York: Universe Books.

Michaelis, L. (2000), "Sustainable consumption and production", in F. Dodds (ed.), *Earth Summit 2002*, London: Earthscan, pp. 264–78.

Mitchell, T. and Maxwell, S. (2010), *Defining Climate Compatible Development*,

London: Climate & Development Knowledge Network, available at: www.cdkn.org/wp-content/uploads/2011/02/CDKN-CCD-DIGI-MASTER-19NOV.pdf (accessed 13 March 2016).

Mohan, G. and Holland, J. (2001), "Human rights and development in Africa: moral intrusion or empowering opportunity?", *Review of African Political Economy*, 28(88), 177–96.

Mol, A., Sonnenfeld, D. and Spaargaren, G. (eds) (2009), *The Ecological Modernisation Reader: Environmental Reform in Theory and Practice*, London and New York: Routledge.

Nilsson, M. and Eckerberg, K. (2007), *Environmental Policy Integration in Practice: Shaping Institutions for Learning*, London: Earthscan.

Nordhaus, W. D. (2007), "A review of *The Stern Review on the Economics of Climate Change*", *Journal of Economic Literature*, 45(3), 686–702.

OECD (Organisation for Economic Cooperation and Development) (2009), *The Economics of Climate Change Mitigation: Policies and Options for Global Action Beyond 2012*, Paris: OECD, available at: www.oecd.org/env/cc/theeconomicsofclimatechangemitigationpoliciesandoptionsforglobalactionbeyond2012.htm (accessed 15 March 2016).

Opschoor, J. B. (2008), "Fighting climate change: human solidarity in a divided world", *Development and Change*, 39(6), 1193–202.

Ravindranath, N. H. and Sathaye, J. (2002), *Climate Change and Developing Countries*, Dordrecht: Kluwer.

Rezai, A., Foley, D. K. and Taylor, L. (2009), "Global warming and externalities", SCEPA Working Paper 2009–3, New School University, New York.

Richardson, K., Steffen, W., Schellnhuber, H. J., Alcamo, J., Barker, T., Kammen, D. M., Leemans, R., Liverman, D., Munasinghe, M., Osman-Elasha, B., Stern, N. and Wæver, O. (2009), *Synthesis Report*, Climate Change: Global Risks, Challenges and Decisions, Copenhagen, 10–12 March, available at: www.pik-potsdam.de/news/press-releases/files/synthesis-report-web.pdf (accessed 15 March 2016).

Rist, G. (2007), "Development as a buzzword", *Development in Practice*, 17(4–5), 485–91.

Sachs, J. (2008), "Technological keys to climate protection", *Scientific American*, 298(4), 40.

SDC (Sustainable Development Commission) (2009), *Prosperity Without Growth*, London: SDC.

SEI (Stockholm Environmental Institute) (2009), *The Economics of Climate Change: Kenya*, Oxford: SEI.

Skea, J. and Nishioka, S. (2008), "Policies and practices for a low-carbon society", *Climate Policy: Supplement – Modelling Long-Term Scenarios for Low-Carbon Societies*, 8(1), 5–16.

Speth, J. G. (2008), *The Bridge at the Edge of the World: Capitalism, the Environment and Crossing from Crisis to Sustainability*, New Haven, CT and London: Yale University Press.

Stern, N. (2006), *Stern Review: The Economics of Climate Change*, London: HM Treasury, available at: http://webarchive.nationalarchives.gov.uk/20100407172811/www.hm-treasury.gov.uk/stern_review_report.htm (accessed 15 March 2016).

Stern, N. (2007), *The Economics of Climate Change*, Cambridge: Cambridge University Press.

Stern, N. (2009), *A Blueprint for a Safer Planet: How to Manage Climate Change and Create a New Era of Progress and Prosperity*, London: Bodley Head.

Storm, S. (2009), "Capitalism and climate change: can the invisible hand adjust the natural thermostat?", *Development and Change*, 40(6), 1011–38.

TEEB (The Economics of Ecosystems & Biodiversity) (2010), *Mainstreaming the Economics of Nature: A Synthesis of the Approach, Conclusions and Recommendations of TEEB*, Geneva: TEEB, available at: www.teebweb.org/publication/mainstreaming-the-economics-of-nature-a-synthesis-of-the-approach-conclusions-and-recommendations-of-teeb/ (accessed 15 March 2016).

Tyndall Centre (2009), "Climate change in a myopic world", Tyndall Briefing Note No. 36, Norwich: Tyndall Centre, available at: www.tyndall.ac.uk/Tyndall-Publications/Briefing-Notes/2009/Climate-change-myopic-world (accessed 18 October 2012).

UNCTAD (2007), *The Least Developed Country Report 2007*, Geneva: UNCTAD.

UNEP (United Nations Environmental Programme) (2011), *Towards a Green Economy: Pathways to Sustainable Development and Poverty Reduction*, Nairobi: UNEP.

UNFCCC (United Nations Framework Convention on Climate Change) (1992), *The United Nations Framework Convention on Climate Change*, Bonn: UNFCCC, available at: http://unfccc.int/resource/docs/convkp/conveng.pdf (accessed 18 October 2012).

UNFCCC (United Nations Framework Convention on Climate Change) (1997), *The Kyoto Protocol*, Bonn: UNFCCC, available at: http://unfccc.int/resource/docs/convkp/kpeng.pdf (accessed 18 October 2012).

UNFCCC (United Nations Framework Convention on Climate Change) (2007), *Bali Action Plan*, Bonn: UNFCCC, available at: http://unfccc.int/resource/docs/2007/cop13/eng/06a01.pdf#page=3 (accessed 18 October 2012).

UNFCCC (United Nations Framework Convention on Climate Change) (2009), *Copenhagen Accord*, Bonn: UNFCCC, available at: http://unfccc.int/resource/docs/2009/cop15/eng/11a01.pdf (accessed 18 October 2012).

UNFCCC (United Nations Framework Convention on Climate Change) (2010), *Cancun Agreements*, Bonn: UNFCCC, available at: http://unfccc.int/resource/docs/2010/cop16/eng/07a01.pdf#page=2 (accessed 18 October 2012).

UNFCCC (United Nations Framework Convention on Climate Change) (2011), *Durban Platform for Enhanced Action*, Bonn: UNFCCC, available at: http://unfccc.int/files/meetings/durban_nov_2011/decisions/application/pdf/cop17_durbanplatform.pdf (accessed 18 October 2012).

Urban, F. (2009), "Sustainable energy for developing countries: modelling transition to renewable and clean energy in rapidly developing countries", PhD thesis, Groningen: University of Groningen.

Urban, F. (2010), "Pro-poor low carbon development and the role of growth", *International Journal of Green Economics*, 4(1), 82–93.

Urban, F. (2011), *Technology, Trade and Climate Policy: The Pursuit of Low Carbon Development in Least Developed Countries, Vulnerable Economies and Small Island Developing States*, London: Commonwealth Secretariat.

Urban, F. (2014), *Low Carbon Transitions for Developing Countries*, London: Routledge.

Urban, F. and Nordensvärd, J. (2013), *Low Carbon Development: Key Issues*, London: Routledge.

Urban, F., Mitchell, T. and Silva Villanueva, P. (2011), "Issues at the interface of disaster risk management and low-carbon development", *Climate and Development*, 3(3), 259–79.

Victor, P. (2008), *Managing Without Growth: Slower by Design, not Disaster*, Cheltenham: Edward Elgar.

Wagner, A. (2004), "Redefining citizenship for the 21st century: from the National Welfare State to the UN Global Compact", *International Journal of Social Welfare*, 13(4), 278–86.

WWF (World Wide Fund for Nature) (1998), *Living Planet Report*, Godalming: WWF.

2 Theoretical framework

Johan Nordensvärd

A brief introduction

Low-carbon development should not be seen as neutral or unproblematic. The whole discourse contains some troublesome issues that concern both the ethical and the moral understanding of development. It is sometimes seen as ironic that low-carbon development often affects poorer countries disproportionately and sometimes implies the imposition of greenhouse gas mitigation on developing countries that have historically done little to cause the climate change problem (Najam, 2005; Page, 2006; Roberts and Parks, 2007; Barker *et al.*, 2008). This becomes even more evident for low-income countries that have a legacy of low emissions and are also unlikely to make a substantial contribution to global emissions rises in the near future. It is therefore a concern that many low-income countries have weaker positions in the international negotiations on tackling climate change, with this procedural inequity creating a barrier to distributive equity (Najam, 2005). The problem of equity (across social groups living today and across generations) raised by climate change, and the need for urgent and extensive mitigation, are ethical problems, and should be informed by moral philosophy (drawing on scientific findings with respect to climate change impacts) and not just by economics in isolation (Barker *et al.*, 2008: 317–18).

Low-carbon development might bring opportunities and benefits for both developed and developing countries; nevertheless, low-carbon development often implies balancing political, economic, social and technological considerations. It is important to understand that low-carbon development is no silver bullet and many of the pre-existent societal problems will be reflected in how low-carbon development is implemented. Social and economic risks need to be carefully analysed and reduced. Low-carbon development lacks a distinctive analytical framework for its social implications.

The mainstream focus lies on economic development that is predominantly low carbon, while poverty reduction, more equal access to environmental goods and more equal exposure to environmental bads take a back seat. As Meg Huby (2001) has discussed, one needs to understand that environmental policies in general cannot be assumed to have only positive impacts; they could also have negative short-term costs for the poorest and most marginalised. This monograph will be a first attempt to look at these negative consequences from a conceptual perspective. This chapter aims to discuss and develop a theoretical framework for understanding the social implications of low-carbon development.

Low-carbon development will mean different priorities in developed and developing countries, with poverty reduction and adaptation more important in poorer countries. More low-carbon development might even mean a reduction in consumption and living standards in the developed countries. It is important to debate the social impacts that low-carbon development can have.

Ironically, there is a risk that low-carbon development might be implemented at the expense of the poorest people in society. This could be the poor in developed countries, who may face higher energy bills and fuel poverty due to utility companies that pass the costs of renewable energy investments onto customers. This could be the poor in developing countries, who may be faced with higher food prices due to the limited availability of land as a result of extensive biofuel developments or who might even be evicted from their land due to biofuel developments or the construction of large dams and who face a desolate future. Low-carbon development therefore needs to be achieved in a way that promotes social and economic benefits for all strata of society, particularly for the poorest.

This book aims to discuss the social implications of low-carbon development and how to face these challenges in a socially just way. It is therefore important to create a theoretical and conceptual framework to explore, critically analyse and propose different ways in which low-carbon development could be understood in both developed and developing countries. Chapter 1 aimed to discuss the academic discourse around low-carbon development; this chapter aims to discuss how one could academically make sense of the social dimensions of low-carbon development.

The dominant understanding of low-carbon development has been and often still is narrowly focused on technological innovation and economic growth. Social dimensions or even social analyses of low-carbon development have been rather lacklustre. To analyse the social implications, it is important to assess how socially sustainable low-carbon development is, whether its social implications could be understood to be just and how these social implications and stakeholders' well-being are the target of relevant social policies.

This chapter will first discuss how the social implications of low-carbon development could be understood, analysed and discussed from the academic/theoretical perspectives of social policy, social sustainability and environmental justice. The selected case studies in this book will illustrate that low-carbon development does not solve underlying social inequalities but does change and morph according to these so that low-carbon development can have both short- and long-term consequences for least well-off.

Social policy and the environment

Social policy has become more and more linked up with the environment since the end of the 1990s. Social policy deals with actions to promote human well-being and the study of it (Alcock, 1997), as well as the academic study of social services and the welfare state (Spicker, 1995). In addition, social policy is considered to be the public management of social risks (Esping-Andersen, 1999: 36); such risks are affected by environmental risks, such as climate change. Social policy as both the activities of the welfare state and an academic subject is often centred in and around the developed countries (such as Sweden, the UK, Germany and the US), which possess government policies and institutions that implement social welfare. Social policy is predominantly understood through a social discourse that is often reliant upon a notion of social citizenship. T. H. Marshall used the discourse of social citizenship to describe European welfare state developments in the middle of the twentieth century. During this period, 'rights' developed to grant working people a modicum of economic welfare, social security and "the right to participate in full in the heritage and economic wealth of society" (Wagner, 2004: 280). This was partially to guarantee the working class a certain standard of living independent of the market. Social rights offered some protection for workers against particular aspects and outcomes of the market such as unemployment (Wagner, 2004: 280). The developed world is predominantly responsible for an over-consumption of the planet's natural resources. A social liberal perspective on consumption has been an underlying assumption in our perception of social citizenship. In Marshall's classical conception, citizens were "first and foremost private individuals and consumers whose freedom of choice had to be protected against government interference" (Wagner, 2004: 280).

In many developing countries, there is no welfare state, which makes people reliant upon informal welfare arrangements. In their seminal work on welfare regimes, Geof Wood and Ian Gough have discussed the concepts of 'informal security regimes' and 'insecurity regimes' (Wood and Gough, 2006), which describe low- and middle-income countries in which the state cannot offer comprehensive or even basic welfare services. Many developing countries have an informal welfare system that relies more upon informal communities and informal social services than upon the state. Wood and Gough discuss the nature of informal security arrangements, whereby rights and duties are often informal and people rely heavily upon community and family relationships to meet their security needs: "rights and entitlements may also be found ... in the informal domains of social relationships and cultural expectations" which in some cases could be "personalized in a range of clientelist and reciprocal (perhaps kin) arrangements" (Wood and Gough, 2006: 1698). Wood and Gough argue that these relationships are often hierarchical and asymmetrical and they mention that "formal security" is the "most satisfactory way of meeting universal human needs including those for security" (Wood and Gough, 2006: 1709).

The distinction between formal and informal welfare has a direct impact on how the poorest can mitigate the costs from both climate change and low-carbon development policies. Gough (2011) suggests that, for developed countries, the distributive consequences of climate change mitigation programmes action will create new social injustices and impose new demands on the welfare state. The same will also be true for developing countries, in which social protection schemes are often limited in both availability and coverage.

The nexus between social policy and the environment has grown in importance since the end of the 1980s and has attracted discussion on the greening of social policy (Cahill, 1991) and the balance between ecological and social rationality (Ferris, 1993). There is therefore a strong rationale for addressing environmental issues when it comes to social policy (Huby, 1998; George and Wilding, 1999; Cahill, 2001). There is an increasing interest in linking up sustainability, low-carbon development and climate change mitigation with social policy (Huby, 1998; Cahill, 2001; Roberts and Parks, 2007; Gough, 2011; Nordensvärd, 2013). The excessive consumption of natural resources, particularly in developed countries and emerging economies, coupled with a rapidly growing global population, poses global challenges both for social policy and low-carbon development. Tim Jackson therefore suggests that "[t]here is no credible, socially just, ecologically sustainable scenario of continually growing incomes for a world of nine billion people" (Jackson, 2009: 57). The conflict is that on the one hand we have rich countries that consume far too many resources and on the other we have the rest of the world dreaming of modernising, which means consuming as much as the developed countries. Benjamin Barber expresses this ambivalence: "Yet this ecological consciousness has meant not only greater awareness but also greater inequality, as modernized nations try to slam the door behind them, saying to developing nations, 'The world cannot afford your modernization; ours has wrung it dry'" (Barber, 1992). One could therefore argue that social policy should aim to protect both humans and the environment; however, this would require reducing the over-consumption of natural resources while trying to limit excessive population growth.

The dominant social policy discussion has often been focused on protecting the living standards of most citizens from the volatile market and more rarely on the need to protect the environment from the over-consumption that is a consequence of living standards in developed countries. This has changed since the rise of an environmental discourse within the fields of both social policy and international development, as more countries aspire to the wealth of those that are already industrialised. Tony Fitzpatrick (2011) highlights that social policies and welfare institutions have been intertwined with industrialism, over-consumption and unsustainable economic growth. Environmental concerns have not been a focus of social policy. Low-carbon development will result in costs that need to be shared equally and the role of social policy is important in mitigating the burden of low-carbon development. Kevin Murphy (2012) states that it is important to connect the social imperatives with the environmental imperatives. Two such environmental imperatives that link the social to the environmental in both developed and developing countries, sustainability and environmental justice, are discussed below.

Sustainability

The discourse of sustainability has become an increasingly important lens through which to understand development projects in general as it combines both environmental and economic factors. The sustainable development and environmental management discourse tries to bridge tensions between humans and the natural environment in which humans live. Kate Kearins and colleagues argue that sustainability is "a systems concept that has at its heart ecological sustainability and the longevity of biophysical systems that support human life" (Kearins *et al.*, 2010: 519). Sustainability was popularised by the Brundtland Report in 1987, which became synonymous with "development that meets the needs of the present without compromising the ability of future generations to meet their own needs" (UN WCED, 1987: 43). There was an interest in integrating the social, economic and environmental dimensions of development. These dimensions were seen as the core pillars of sustainable development. Even though sustainability has become one of the most important environmental concepts in development, the social pillar was often seen as less of a priority than both the environmental and economical pillars. For a long time, the social pillar was also seen as being vague and marginal (Dempsey *et al.* 2011; Casula Vifell and Soneryd, 2012). Some argue that the social pillar is the most elusive aspect of sustainable development (Thin, 2002).

The original Brundtland Report did highlight the social pillar relatively strongly and argued that it was important that the "the essential needs of the world's poor" should be given an "overriding priority" (UN WCED, 1987: 43). At the same time, it was seen that the state of technology and social organisation had a direct impact on how the environment could meet the needs of the poorest (UN WCED, 1987: 43/37). Thus, to be able to meet the needs of the poorest, it was paramount first to develop both technology and social organisation. This implied a lesser focus on the social pillar, which was often seen as a positive by-product. We should therefore be under no illusion that the original meaning of sustainability was very much linked to economic growth. The Brundtland Report argued that as advances in technology and social organisation are achieved, this will spark a new era of economic growth: "[If] large parts of the developing world are to avert economic, social and environmental catastrophes, it is essential that global economic growth be revitalized. In practical terms, this means more rapid economic growth in both industrial and developing countries" (UN WCED, 1987: 89).

This perception of sustainability takes for granted that there is no trade-off between environmental sustainability, economic progress and poverty reduction. This has become the hegemonic understanding of sustainability that is reflected in influential reports such as the UNEP Green Economy (2011). There have been social elements to both the Brundtland Report and the documents resulting from the Rio meetings, but many argue that the social dimension has been secondary to environmental and economic considerations (Marcuse, 1998; Agyeman, 2008; Bebbington and Dillard, 2009). The focus on the economy and growth is very much evident in how the three pillars have been summed up: the three Ps (people, planet and profit) or the three Es (environment, economy and equity) (Boström, 2012: 3). It is taken for granted that humans could achieve a balance between environment, economy and equity without changing the global capitalist framework. There is a belief that sustainable growth and efficient sustainable management techniques will be able to eradicate poverty and improve global equity (Littig and Grießler, 2005). This openness to economic growth and liberalism has at the same time ensured that both sustainability and low-carbon development have become mainstreamed.

The peak of sustainability was in the 1990s, sparked by the United Nations Conference on Sustainable Development in Rio de Janeiro. Twenty years later, the Rio+20 summit put sustainable development back onto the agenda. This will be followed by the Sustainable Development Goals (SDGs), which will build upon the Millennium Development Goals (MDGs) and be in line with the post-2015 international development agenda. While environmental and economic development will be at the forefront of many of the goals, social goals are also firmly embedded in the SDGs.

The social pillar of sustainability has been relatively neglected compared to the economic and environmental pillars (Cuthill, 2009). Many scholars highlight the importance of developing a greater link between the social and environmental pillars (Dobson, 2003; Littig and Grießler, 2005; Gough *et al.*, 2008). Social sustainability has remained a rather fluid and ambiguous concept that is hard to pin down to a more specific meaning. Nicola Dempsey and colleagues (2011) highlight the open-endedness of the concept and state that social sustainability is a dynamic concept that changes over time; it is neither absolute nor constant. There have been ongoing attempts to link social sustainability to the other dimensions of sustainable development and wider policy issues (Littig and Grießler, 2005;

Davidson, 2009; Dillard *et al.*, 2009; Casula Vifell and Soneryd, 2012; Dempsey *et al.*, 2011). Some scholars highlight that social sustainability lacks a clear and widely accepted basis for analysis and there is no common unit of measurement such as monetary units or CO_2 levels that could be used in research to analyse both economic and environmental sustainability (Bebbington and Dillard, 2009). Kristen Magis and Craig Shinn (2009) attempt to define four universal principles of social sustainability:

1 human well-being
2 equity
3 democratic government
4 democratic civil society.

Murphy (2012) has undertaken an extensive review of the literature related to sustainable development and has suggested four pre-eminent policy concepts for social sustainability:

1 *Equity* is seen as a key social concept and is linked to the distribution of welfare goods and life chances "on the basis of fairness and it applies to national, international and intergenerational contexts" (Murphy, 2012: 20). Murphy argues that equity as a policy concept would mean research covering everything from the export of pollution and how to curb it, how to decarbonise welfare services and how to reduce the current generation's consumption to save resources for coming generations to helping southern countries to cope with climate change through economic transfers (Murphy, 2012: 21).
2 *Awareness for sustainability* refers to activities and processes to raise "public awareness of sustainability issues with a view to encouraging alternative, sustainable consumption patterns" (Murphy, 2012: 23). This often includes education through sustainable development programmes or programmes to inform consumers to make green choices. This could also include education that challenges the dominant perception of growth and promotes non-material happiness (Murphy, 2012: 21).
3 *Participation* refers to the overarching policy goal "of including as many social groups as possible in decision-making processes" and the belief that by "joining in participatory processes, individuals and groups can enhance their social inclusion" (Murphy, 2012: 24). There is an overall argument that an extensive and just participation process will give government policy legitimacy. It is considered important to broaden participation in the environmental planning process to include marginalised and vulnerable groups and to consider future generations (Murphy, 2012: 21).
4 *Social cohesion* refers to a policy discourse that is often used in the EU and that is generally geared towards "promoting happiness/well-being;

minimizing social strife; reducing crime; promoting interpersonal trust; and combating suicide, bullying, and antisocial behaviour" (Murphy, 2012: 24). Within this discourse, it has become a political goal to promote social activities and planning processes that aim for environmental goals. These could include projects such as transitions towns but also include infrastructural projects (Murphy, 2012: 21).

Murphy's comprehensive literature review shows that there is a considerable amount of research that aims to promote how environmental sustainability could or should be equitable. Nevertheless, it is important to acknowledge that social and environmental goals might not be congruent, and that the global social arrangements might need to be reformed as well as our economic arrangements to facilitate environmental sustainability. Moreover, Peter Marcuse argues that sustainability should not mean promoting the current global status quo with all its inequalities (Marcuse, 1998). There have been attempts to link social sustainability to the concept of environmental justice. This makes sense as environmental justice has a more developed social justice perspective compared to the original social pillar in sustainable development. Julian Agyeman and Bob Evans (2004) argue that "just sustainability" needs a clear linkage between sustainable development and environmental justice to prevent the social pillar from becoming one-sided. David Harvey (1996) points out that addressing environmental injustices needs to be a first priority on the sustainability agenda.

Environmental justice

Environmental justice differs from other environmental discourses by defining the environment as the set of linked places "where we live, work and play" (Turner and Wu, 2002: 4). The environmental justice framework originated in a US context that focused around issues of race and ethnicity and how these were intertwined with the distributions of environmental bads, such as pollution and technological risk (Bullard, 1999). Environmental justice became the concept for describing political activism in the US to "resist the imposition of toxic and polluting facilities in minority and poor communities" (Walker, 2009: 356).

Distributive justice became a focal point in framing how communities of colour have been exposed to environmental hazards, such as toxic waste or other environmental issues. Categories such as gender, race and class and the impact of environmental bads and access to environmental goods such as quality of life, natural resources and a clean environment became central to the discussion of distributive justice and the environment (Boström, 2012: 5). The focus on distribution has often been expanded upon with other dimensions, such as participation, recognition and capabilities. Laura Pulido (1996), Daniel Faber (2005) and David Schlosberg

(2007) have highlighted the importance of process and production in environmental justice. There is an overarching perception that most pollution and degradation is caused by the more affluent and powerful, and that the environmental consequences hit the poor disproportionately. It is therefore important that the decision-making processes in environmental policy and specific projects are transparent, just and participative. Distributive justice is concerned with whether environmental resources are allocated in a fair and equitable way among diverse groups of citizens and stakeholders (Maiese, 2003). Distributive justice has often focused on how marginalised groups have been exposed to environmental hazards, such as toxic substances or other environmental issues. The environmental justice movement started in California to struggle against the effects of industrial contamination on "poorer communities of colour" (Lohmann, 2009: 1075). There have been concerns that marginalised groups live in areas that contain environmental hazards such as toxic waste or danger of flooding. Much of the early research in the US has been to discern how to prevent marginalised groups from being faced with environmental risks. Some examples have been focused on pollution treatment in waste collection sites or other industrial activities.

Recent research has expanded on this original rather limited interpretation of environmental justice and the literature has been growing richer and focused more on the distribution of environmental goods and bads (Agyeman and Evans, 2004). The importance lies in creating fair processes for environmental policy-making and policy implementation. There is a perception that if the policy-making and implementation have been fair, participating parties tend to accept a disliked outcome (Deutsch, 2000).

Environmental justice has become politically mainstreamed in the US, which has meant that the Environmental Protection Agency (EPA) "has had an Office of Environmental Justice since 1992 and made efforts to integrate environmental justice into activities across the EPA" (Gunnarsson-Östling and Höjer, 2011: 1050). Moreover, environmental justice is not just an academic field and an policy framework; it is also a social movement, which means that communities engage and react against environmental bads in the community (Agyeman, 2005: 3).

Entitlements "concentrate on setting minimal standards such as a universal right to a clean and healthy environment" and "do not necessarily safeguard a just distribution of environmental goods or bads, but they do provide a minimum standard for all" (Gunnarsson-Östling and Höjer, 2011: 1051). Just like discussion of minimum wage/minimum living standards, a minimal "right to a safe, healthy and environmentally sound environment" has become an major part of the human rights discourse (UNCHR, 1994: 74).

Procedural justice is less about equal distribution and more about the direct empowerment and participation of different stakeholders in environmental processes (Boström, 2012: 5). David Schlosberg (2004) discusses the links between political participation and recognition, drawing upon the discussion by both Iris Young and Nancy Fraser. The overall argument is that one needs to acknowledge that a lack of respect and recognition could lead to a "decline in a person's membership and participation in the greater community, including the political and institutional order" (Schlosberg, 2004: 519). Schlosberg argues that we also need to add a capability dimension to environmental justice; this would "enrich conceptions of environmental and climate justice by bringing recognition to the functioning of these systems, in addition to those who live within and depend on them" (Schlosberg, 2013: 44). This approach has created an attempt to include environmental concerns in an environmental justice movement that has often been perceived to be anthropocentric (Shrader-Frechette, 2002).

> When we interrupt, corrupt, or defile the potential functioning of ecological support systems, we do an injustice not only to human beings, but also to all of those non-humans that depend on the integrity of the system for their own functioning.
>
> (Schlosberg, 2013: 44)

Procedural justice is less about equal distribution and more about the direct empowerment and participation of different stakeholders in environmental processes. Much research therefore delves into investigating whether and to what extent social groups have access to information and modes of participation in environmental issues and environmental planning.

Over the decades, environmental justice has expanded both in theory and in geographical application. The interplay between global and local manifestations and agendas has become more important (Walker and Bulkeley, 2006; Schlosberg, 2007; Carruthers, 2008; Schroeder *et al.*, 2008; Sze and London, 2008). Julie Sze and Jonathan K. London suggest that the environmental justice framework is today used at a global scale, incorporating both global and local concerns on an interdisciplinary basis (Sze and London, 2008). Julian Agyeman and Bob Evans (2004) argue that 'just sustainability' needs a clear linkage between sustainable development and environmental justice to prevent the social pillar from becoming one-sided. David Harvey (1996: 385) points out that environmental injustices need to be a first priority on the sustainability agenda. The importance of these concepts in looking at social impacts tends to focus on "distributive (fair allocation of resources) and especially procedural justice (recognition, participation and power distribution)" (Karjalainen and Järvikoski, 2010).

Analytical framework

In this book we focus on looking at distributive and procedural dimensions as a way to understand how socially sustainable a low-carbon development scheme is (see Box 2.1). We work with Magnus Boström's (2012) conceptualisation of social sustainability and environmental justice. His 'what' category covers *substantive aspects* of social sustainability, hence covering the distributive dimension. His 'how' category covers *procedural aspects* of social sustainability. This book examines how substantive the social sustainability of low-carbon development could be. The book thereby questions how a string of low-carbon schemes have been able to adjust their design and implementation according to environmentally just criteria as defined in this chapter. This book argues that it is important to consider environmental implications, social implications, social sustainability and social policy in a concerted analysis of low-carbon development projects. The book will use Box 2.1 as a way to analyse and discuss particular low-carbon developments.

With substantive dimensions, we have adapted Boström's (2012) categories for analysing the social impacts of the selected low-carbon schemes. 'Substantive dimension' here means researching how the affected people have been able to meet basic needs and have access to basic services, and to analyse whether their burden is just and fair. With procedural dimensions, we have adapted Boström's (2012) categories to understand whether low-carbon schemes have been empowering or disempowering with regard to the affected population. This means explicitly researching whether low-carbon schemes have been accountable in both governing and managing the process. Boström adds that social sustainability needs to incorporate both procedural and substantive dimensions to understand how social sustainability could both include "the improvement of conditions for living people and future generations *and* the quality of governance of the development process" (Boström, 2012: 8).

The second aspect that we look at is the procedural aspect to analyse how accountable the processes are, and whether transparent information, communication and participation is happening in low-carbon development schemes. In this book we will look at the substantive and procedural dimensions of low-carbon schemes in both developing and developed countries (see the case study selection criteria below). Many scholars highlight that the substantive and procedural aspects of social sustainability are interdependent dimensions (Agyeman and Evans, 2004; Dillard *et al.*, 2009; Boström, 2012). This book will first and foremost focus on distributive and procedural justice as a way to analyse low-carbon development and its social implications. This boils down to the fact that the "legitimacy of environmental governance rests on both distributive and procedural justice, and these two are tied together, as unequal distribution of wealth often translates into unequal participation in collective decisions" (Karjalainen and Timo Järvikoski, 2010: 320).

I have chosen five case studies through which to discuss and analyse low-carbon development in both developing and developed countries, covering some of the major trends: hydropower dams, feed-in tariffs, carbon offsetting, low-carbon housing and solar power for refugees.

Box 2.1 Substantive and procedural aspects of social justice

Substantive aspects: what social sustainability goals to achieve?

- Basic needs such as food, housing, and income and extended needs such as recreation and self-fulfillment
- Inter- and intra-generational justice along gender, race, class, and ethnicity dimensions
- Fair distribution of income
- Fair distribution of environmental "bads" and "goods"
- Equality of rights, including human rights, land user and tenure rights, and indigenous people's rights
- Access to social infrastructure, mobility, local services, facilities, green areas, and so forth
- Employment and other work-related issues,facilitating for local small and medium enterprises
- Opportunity for learning and self-development
- Community capacity for the development of civil society and social capital
- Security (e.g. economic, environmental)
- Health effects among workers, consumers, and communities
- Social cohesion, inclusion, and interaction
- Cultural diversity and traditions
- Sense of community attachment, belonging, and identity
- Social recognition
- Attractive housing and public realm
- Quality of life, happiness, and well-being

Procedural aspects: how to achieve sustainable development?

- Access to information about risks and the sustainability project
- Access to participation and decision making in different stages of the process and over time
- Proactive stakeholder communication and consultation throughout the process
- Empowerment for taking part in the process (e.g. awareness, education, networking, economic compensation)
- Participating in the framing of issues, including defining criteria, scope, and subjects of justice
- Social monitoring of the policy, planning, and standard-setting process
- Accountable governance and management of the policy, planning, and standard-setting process

(Boström, 2012: 6)

Case Study 1 (Chapter 3): The social challenges of Chinese hydropower dams in the Mekong region

This chapter discusses the social implications of Chinese-funded and Chinese-built hydropower dams overseas. Due to China's rapid economic growth, its rapid industrialisation and its limited domestic natural resources, the Chinese government has issued the 'Going Out Strategy', which promotes investments in overseas natural resources like water and energy. In the search for climate-friendly low-carbon energy, cheap electricity and access to growing markets, profits and employment overseas, Chinese institutions have turned to Southeast Asia, where they are currently involved in numerous ongoing large hydropower projects as contractors, investors, regulators and financiers. These Chinese institutions have an influence on the environment and social practices as well as on diplomatic and trade relations in the host countries. This chapter explores the social implications of these activities, such as the displacement and resettlement and the declines in livelihoods of people affected by the dam sites. The chapter also explores the conflict that arises at the sites of the dams and suggests how hydropower dams could be built in the future with fewer social ramifications.

Case Study 2 (Chapter 4): The high costs of wind energy in Germany: social challenges and possible solutions

This chapter discusses wind energy policy and the social challenges that energy prices might entail. This chapter also discusses what this means for the way in which low-carbon transitions should be financed and organised. The German energy transition has often been portrayed as being a pioneer for investments in renewable energy such as wind power. The growth of wind energy in the last two decades in Germany has been very dependent upon governmental decisions for favourable regulations and feed-in tariffs. The feed-in tariff for wind energy has led to an upscaling of wind energy turbine sizes and capacities as well as an increase in investment in riskier and more costly offshore wind energy farms. There are also major issues due to a lack of investment in the grid system. The demand for wind energy cannot keep up with the supply due to bottlenecks in the grid system, particularly regarding the distribution of wind energy from the windy North to the South. The social implications have included an increase in costs; in addition, the wind energy feed-in tariff has promoted large-scale investments supporting corporate multinational actors, which has meant a relatively limited focus on smaller-scale wind power, community participation and the involvement of citizens. Most important is the fact that consumers have paid for the largest part of the energy transition through rising energy costs while corporations have been reaping profits and energy-intensive corporations have received favourable financial treatment.

Case Study 3 (Chapter 5): Social implications of carbon markets: the case of carbon offsets and Plantar in Brazil

This case study discusses the social implications of large-scale industrial tree plantations in Brazil and the problems associated with using carbon markets as a mitigation strategy for climate change. Brazil, like the other BRICS countries, has seen rapid industrial growth over the last 15 years. However, Brazil's industrial expansion is intrinsically linked to its abundance of natural resources, large-scale agro-industry and the recent offshore pre-salt petroleum industry. Since the introduction of carbon trading, Brazil has been a top seller of offset credits. Carbon trading is a primary mitigation scheme used to address climate change. Proponents argue that setting a price and trading carbon dioxide equivalent (CO_2e) emissions is a cost-effective way to reduce pollution. This chapter aims to argue the limits of carbon markets and discuss them from a social justice perspective. In addition, the chapter explores the social implications of 'reforestation' (i.e. tree plantations) as a way to offset the carbon emissions from polluting industries and discusses some of the most critical challenges of these projects. We will look at a case study in Minas Gerais, Brazil, where the World Bank continues a carbon emissions offsetting project with monocultural and non-native eucalyptus plantations.

Case Study 4 (Chapter 6): Domestic energy efficiency policy and fuel poverty in England

Attempting to address both social and environmental dimensions of domestic energy has been described by some as a 'wicked' policy problem given the potential complexities of pursuing two very different sets of policy goals simultaneously. In England, given the poor quality of much of the housing stock, one effective low-carbon, anti-fuel-poverty measure is energy efficiency improvements within the homes of those considered most vulnerable to fuel poverty. Domestic energy policy has typically (at least in part) taken this approach. This chapter first outlines the social and environmental aims implicit in most domestic energy policies and the challenges that policy-makers are faced with in terms of balancing these. The chapter then introduces the two main energy efficiency programmes developed by the UK's Conservative–Liberal Coalition government (2010–15) and implemented in England, and critically analyses their environmental and social justice implications. The chapter concludes by arguing that the Coalition's policy has failed to pursue either environmental or social goals sufficiently.

Case Study 5 (Chapter 7): Pro-poor technology transfer: the case of solar power and refugees in Africa

While the technology transfer of low-carbon energy technology has reached predominantly emerging economies such as China and India, there has been little progress for poorer countries. This chapter looks at attempts to bring low-carbon energy technology to the poorest and most vulnerable people in Africa. The Darfur Solar Cooker Project provided solar cookers to Darfuri refugee women and girls in refugee camps in Chad as an alternative to using fuelwood. The project was driven by a consortium of NGOs, comprising TchadSolaire, the Christian Outreach Relief Development (CORD), Solar Cookers International, the Dutch aid organisation KoZon and Jewish World Watch (JWW), operating in refugee camps run by the United Nations High Commission for Refugees (UNHCR). This is an example of how low-carbon energy can help to improve people's lives as well as reduce pressure on the environment in a context of political instability. Instead of women and girls being forced to leave the safety of their refugee camps in Chad and risk being assaulted in their pursuit of fuelwood, the solar cookers reduced (or even eliminated) the need for fuelwood and thereby allowed women and girls to stay safely in the camp (Urban and Lind, 2011). This chapter aims to discuss whether providing solar energy products such as solar cookers for refugees could be a start in making low-carbon technology transfer more socially just and more accessible to the world's poorest and most disadvantaged. At the same time, the chapter raises criticisms, particularly with regard to the temporary and ad hoc way in which these emergency relief activities are organised.

References

Agyeman, J. (2005), *Sustainable Communities and the Challenge of Environmental Justice*, New York: New York University Press.

Agyeman, J. (2008), "Toward a 'just' sustainability?", *Continuum: Journal of Media & Cultural Studies*, 22(6), 751–6.

Agyeman, J. and Evans, B. (2004), "'Just sustainability': the emerging discourse of environmental justice in Britain?', *Geographical Journal*, 170(2), 155–64.

Alcock, P. (1997), "The subject of social policy", in P. Alcock, A. Erskinem and M. May (eds), *The Student's Companion to Social Policy*, Oxford: Blackwell.

Barber, B. (1992), "Jihad Vs McWorld", *Atlantic Monthly*, March, available at: www.theatlantic.com/past/politics/foreign/barjiha.htm (accessed 30 April 2016).

Barker T., Scrieciu, S. and Taylor, D. (2008), "Climate change, social justice and development", *Development: Journal of the Society for International Development*, 51(3), 317–24.

Bebbington, J. and Dillard, J. (2009), "Social sustainability: an organizational-level analysis", in J. Dillard, V. Dujon and M. King (eds), *Understanding the Social Dimension of Sustainability*, New York: Routledge, pp. 157–73.

Boström, M. (2012), "A missing pillar? Challenges in theorizing and practicing social sustainability: introduction to the special issue", *Sustainability; Science, Practice and Policy*, 8(12), 3–15.

Bullard, R. D. (1999), "Dismantling environmental racism in the USA", *Local Environment*, 4(1), 5–19.

Cahill, M. (1991), "The greening of social policy?", in N. Manning (ed.), *Social Policy Review 1990–91*, New York: Longman, pp. 9–23.

Cahill, M. (2001), *The Environment and Social Policy*, London: Routledge.

Carruthers, D. V. (2008), "The globalization of environmental justice: lessons from the U.S.–Mexico border", *Society and Natural Resources*, 21(7), 556–68.

Casula Vifell, Å. and Soneryd, L. (2012), "Organizing matters: how the 'social dimension' gets lost in sustainability projects", *Sustainable Development*, 20(1), 18–27.

Cuthill, M. (2009), "Strengthening the 'social' in sustainable development: developing a conceptual framework for social sustainability in a rapid urban growth region in Australia", *Sustainable Development*, 18(6), 362–73.

Davidson, M. (2009), "Social sustainability: a potential for politics?", *Local Environment*, 14(7), 607–19.

Dempsey, N., Bramley, G., Power, S. and Brown, C. (2011), "The social dimension of sustainable development: defining urban social sustainability", *Sustainable Development*, 19(5). 289–300.

Deutsch, M. (2000), "Justice and conflict", in M. Deutsch and P. T. Coleman (eds), *The Handbook of Conflict Resolution: Theory and Practice*, San Francisco, CA: Jossey-Bass, pp. 45–65.

Dillard, J., Dujon, V. and King, M. (eds) (2009), *Understanding the Social Dimension of Sustainability*, New York: Routledge.

Dobson, A. (2003), *Citizenship and the Environment*, New York: Oxford University Press.

Esping-Andersen, G. (1999), *Social Foundations of Postindustrial Economies*, Oxford: Oxford University Press.

Faber, D. R. (2005), "Building a transnational environmental justice movement: obstacles and opportunities in the age of globalization", in J. Bandy and J. Smith (eds), *Coalitions Across Borders: Transnational Protest and the Neoliberal Order*, Lanham, MD: Rowman & Littlefield, pp. 43–68.

Ferris, J. (1993), "Ecological versus social rationality: can there be green social policies?", in A. Dobson and P. Loucardie (eds), *The Politics of Nature*, London: Routledge, pp. 145–60..

Fitzpatrick, T. (2011), "Challenges for social policy", in T. Fitzpatrick (ed.), *Understanding the Environment and Social Policy*, Bristol: Policy Press, pp. 61–90.

George, V. and Wilding, P. (1999), *British Society and Social Welfare: Towards a Sustainable Society*, New York: St. Martin's Press.

Gough, I. (2011), "Climate change, double injustice and social policy: a case study of the United Kingdom", UNRISD Occasional Paper 1: Social Dimensions of Green Economy and Sustainable Development, United Nations Research Institute for Social Development, Geneva.

Gough, I., Meadowcroft, J., Dryzek, J., Gerhards, J., Lengfeld, H., Marandya, A. and Ortiz, R. (2008), "JESP symposium: climate change and social policy", *Journal of European Social Policy*, 18(4), 325–44.

Gunnarsson-Östling, U. and Höjer, M. (2011), "Scenario planning for sustainability in Stockholm, Sweden: environmental justice considerations", *International Journal of Urban and Regional Research*, 35(5), 1048–67.

Harvey, D. (1996), *Justice, Nature and the Geography of Difference*, Oxford: Blackwell.

Huby, M. (1998), *Social Policy and the Environment*, Buckingham: Open University Press.

Huby, M. (2001), "The sustainable use of resources on a global scale", *Social Policy and Administration*, 35(5), 521–37.

Jackson, T. (2009), *Prosperity Without Growth*, London: Earthscan.

Karjalainen, T. P. and Järvikoski, T. (2010), "Negotiating river ecosystems: impact assessment and conflict mediation in the cases of hydro-power construction", *Environmental Impact Assessment Review*, 30(5), 319–27.

Kearins, K., Collins, E. and Tregidga, H. (2010), "Beyond corporate environmental management to a consideration of nature in visionary small enterprise", *Business Society*, 49(3), 512–47.

Littig, B. and Grießler, E. (2005), "Social sustainability: a catchword between political pragmatism and social theory", *International Journal of Sustainable Development*, 8(1–2), 65–79.

Lohmann, L. (2009), "Climate as investment", *Development and Change*, 40(6), 1063–83.

Magis, K. and Shinn, C. (2009), "Emergent principles of social sustainability", in J. Dillard, V. Dujon and M. King (eds), *Understanding the Social Dimension of Sustainability*, New York: Routledge, pp. 15–44.

Maiese, M. (2003), "Distributive justice", *Beyond Intractability*, G. Burgess and H. Burgess (eds), Conflict Research Consortium, University of Colorado, Boulder, CO, available at: www.beyondintractability.org/essay/distributive_justice/ (accessed 16 March 2016).

Marcuse, P. (1998), "Sustainability is not enough", *Environment and Urbanization*, 10(2), 103–111.

Murphy, K. (2012), "The social pillar of sustainable development: a literature review and framework for policy analysis", *Sustainability; Science, Practice and Policy*, 8(12), 15–30.

Najam, A. (2005), "Developing countries and global environmental governance: from contestation to participation to engagement", *International Environmental Agreements*, 5(3), 303–21.

Nordensvärd, J. (2013), "Social policy and low carbon development", in F. Urban and J. Nordensvärd (eds), *Low Carbon Development: Key Issues*, London: Routledge, pp. 66–79.

Page, E. (2006), *Climate Change and Future Generations*, Cheltenham: Edward Elgar.

Pulido, L. (1996), *Environmentalism and Economic Justice: Two Chicano Struggles in the Southwest*, Tucson, AZ: University of Arizona Press.

Roberts, J. T. and Parks, B. C. (2007), *A Climate of Injustice: Global Inequality, North–South Politics and Climate Policy*, Cambridge, MA: MIT Press.

Schlosberg, D. (2004), "Reconceiving environmental justice: global movements and political theories", *Environmental Politics*, 13(3), 517–40.

Schlosberg, D. (2007), *Defining Environmental Justice*. Oxford: Oxford University Press.

Schlosberg, D. (2013), "Theorising environmental justice: the expanding sphere of a discourse", *Environmental Politics*, 22(1), 37–55.

Schroeder, R., Martin, K. S., Wilson, B. and Sen, D. (2008), "Third World environmental justice", *Society & Natural Resources*, 21(7): 547–55.

Shrader-Frechette, K. (2002), *Environmental Justice: Creating Equality, Reclaiming Democracy*, New York: Oxford University Press.

Spicker, P. (1995), *Social Policy: Themes and Approaches*, New Jersey: Prentice Hall.

Sze, J. and London, J. K. (2008), 'Environmental justice at the crossroads", *Sociology Compass*, 2(4), 1331–54.

Thin, N. (2002), *Social Progress and Sustainable Development*, London: ITDG Publishing.

Turner, R. L. and Wu, D. P. (2002), *Environmental Justice and Environmental Racism: An Annotated Bibliography and General Overview, Focusing on U.S. Literature, 1996–2002*, Department of Environmental Science, Policy and Management, University of California, Berkeley.

UNCHR (UN Commission on Human Rights) (1994), Sub Commission on Prevention of Discrimination and Protection of Minorities, Human Rights and the Environment, *Final Report of the Special Rapporteur*, Geneva: UNCHR.

UNEP (United Nations Environmental Programme) (2011), *Towards a Green Economy: Pathways to Sustainable Development and Poverty Reduction*, Nairobi: UNEP.

UN WCED (World Commission on Environment and Development) (1987), *Our Common Future*, Oxford: Oxford University Press.

Urban, F. and Lind, J. (2011), "Low carbon energy and conflict: a new agenda", *Boiling Point*, 59, available at: www.hedon.info/BP59_Low+carbon+energy+and+conflict (accessed 27 March 2016).

Wagner A. (2004), "Redefining citizenship for the 21st century: from the National Welfare State to the UN Global Compact", *International Journal of Social Welfare*, 13(4), 278–86.

Walker, G. (2009), "Globalizing environmental justice", *Global Social Policy*, 9(3), 355–82.

Walker, G. and Bulkeley, H. (2006), "Geographies of environmental justice", *Geoforum* 37(5), 655–9.

Wood, G. and Gough, I. (2006), "A comparative welfare regime approach to global social policy", *World Development*, 34(10), 1696–712.

3 The social challenges of Chinese hydropower dams in the Mekong region

Johan Nordensvärd with Frauke Urban

Hydropower and the social implications

Hydropower dams are symbols of economic development in general and low-carbon development in particular. Often hydropower is seen as a way in which humans can tame and harness the power of nature for our benefit through development and technology. Very few practical examples of low-carbon development symbolise the power of nature more than the enormous dams that result in the flooding of valleys and towns and change whole ecological systems. At the same time, hydropower dams are controversial for the same reasons, as humans and nature have to make way for these mega-projects. The focus on large-scale technological solutions often comes at the price of neglecting some of their most critical social implications (Urban *et al.*, 2015). Large hydropower dams have been the subject of controversy and debate for several decades due to their large-scale and often irreversible social and environmental impacts (WCD, 2000).

Many large hydropower dams have been situated in sensitive and environmentally important places, which has prompted opposition from local people. An example of this could be the hydropower dams in Tasmania, Australia, which resulted in both controversy and the birth of the Australian green movement to protect the environment. One of the first dam projects was situated in the Lake Pedder National Park and in fact consisted of three dams (the Serpentine Dam, the Scotts Peak Dam and the Edgar Dam); the natural lake was enlarged and became Australia's largest freshwater lake. The environmental consequences gave rise to protests that were unable to stop or adjust the project, which led to some species becoming extinct. The proposed Franklin Dam, which was also due to be situated in the Tasmanian wilderness, generated far more political and social opposition, which led to the indefinite shelving of the hydropower dam project. The list of controversial hydropower dams in sensitive social and environmental areas is becoming longer with the years. Especially controversial are those hydropower dams in developing countries that have had even more far-reaching social and environmental consequences.

Recently Chinese financiers such as the Export Import (ExIm) Bank and other banks have become the world's biggest funders of large hydropower dams. This position was formerly occupied by the World Bank and other regional banks. These large hydropower dams have clearly demonstrated that such projects are deeply problematic. Particularly controversial projects funded by the World Bank led to public critique of hydropower investments. The World Bank was "accused of funding projects without considering the environmental devastation, social disruption, or negative impacts on the economies of the developing nations it intended to aid" (MacDonald, 2001: 1013). The controversial nature of hydropower projects such as the Sardar Sarovar Dam projects in India, the Chonoy Dam in Guatemala and the Itaparica hydropower scheme in Brazil forced the World Bank to introduce safeguard policies in the 1980s to protect local populations from the consequences of dams. Safeguard policies exist with regard to issues such as "environmental assessment, natural habitats, pest management, compensation for involuntary resettlement, indigenous peoples, forests, safety of dams, cultural property, projects in international waterways and projects in disputed areas" (Hall, 2007: 170).

During the reorganisation in 1987, the World Bank decided to set up Environmental Units in all four of their regions, consisting of social scientists as well as environmentalists. Anthony Hall argues this should not be understood as evidence of a growing environmental concern of the bank at the time but instead as

> [a direct] response to widespread adverse publicity received by the Bank within the USA and globally at the hands of a vigorous non-governmental organization (NGO) and media campaign following the catastrophic impacts of major projects, in particular the Northwest Amazon settlement project (Polonoroeste) in Brazil, among others, during the early 1980s.
>
> (Hall, 2007: 161–2)

One example of such a campaign was the '50 Years Is Enough' campaign that demanded that the World Bank and IMF be reformed (Hall, 1994). Hall argues that the US Congress and US Treasury Department played a strong role in bringing about these changes: "[T]he US Treasury under James Baker insisted that the Bank should 'clean up its act' as a precondition for approval of an International Bank for Reconstruction and Development (IBRD) capital increase and IDA replenishment" (Hall, 2007: 171).

The negative experiences of World Bank projects led to the formation of World Commission on Dams (WCD), which aims to research the environmental, social and economic impacts of hydropower dams globally. Its existence is a testament to the difficulty of creating socially sustainable hydropower dams. The WCD highlights that the "overall magnitude, extent and complexity of adverse social impacts for the displaced and dependants on the riverine ecosystem (...) are often not acknowledged and considered in project planning and operations" (WCD, 2000: 98). In 2000, the WCD attempted to develop a framework entitled *Dams and Development: A New Framework for Decision-Making* that could facilitate the planning, implementation and operation of dams. The International Hydropower Association (IHA) launched its own sustainability guidelines in 2004, which were followed by its Sustainability Assessment Protocol in 2006. There has also been a rise in assessment techniques such as Social Impact Assessment, multi-stakeholder platforms and Transboundary Environmental Impact Assessments (Mirumachi and Torriti, 2012).

The World Bank had withdrawn itself from funding large hydropower dams and has only recently re-engaged with hydropower funding through projects such as the Nam Theun 2 (NT2) hydropower project in central Laos. NT2 is seen as a flagship project; however, it still evokes critiques that large hydropower dams are hardly environmentally and socially sustainable. The recurring negative environmental and social implications of large hydropower dams are presented in Table 3.1.

Chinese-funded and Chinese-built large hydropower dams

The aim of this chapter is to discern the social challenges posed by Chinese-funded and Chinese-built large hydropower dams in low- and middle-income countries in the Mekong region in general and in Cambodia in particular. China in general and the ExIm Bank/Sinohydro in particular have become the largest funders and builders of hydropower projects worldwide. This should be linked with China's rather dramatic assumption of the role of the world's leading manufacturer. China's rapid development over the last three decades is unprecedented (Mohan and Power, 2008). In this period, China has witnessed an average annual GDP growth of about 9 per cent through its "state-orchestrated market" approach (Ampiah and Naidu, 2008: 330). However, the high economic growth, rapid modernisation and industrialisation have taken their toll on the Chinese environment. China faces significant environmental challenges such as climate change, resource scarcity and pollution of soil, air and water. China's rapidly growing economy, the high amount of embodied carbon emissions due to its exports and its strong dependence upon coal have made the country a major polluter (Watson and Wang, 2007).

'Fuelling the dragon' has become a major challenge as China's growing economy and its large population can only be sustained using enormous energy resources. This has become increasingly problematic since China is relatively poor in resources and relies heavily upon coal as a dominant energy source. China itself has access to its own sources of coal that is cheap but mostly of low quality. This is supplemented by importing higher-quality coal from other middle- and low-income countries, such as South Africa, and oil from various countries in Africa, Asia and Latin America.

Table 3.1 Environmental and social implications of hydropower

Environmental implications	*Social implications*	*Social sustainability*	*Social policy*
Direct negative environmental impacts include increased erosion, increased sedimentation rates, increased frequency of landslides, changes in water flows, the destruction of flora and fauna, ecosystem changes, geomorphologic changes, decreases in water quality partly due to increased inflows of pesticides and industrial waste waters, increased eutrophication (Chang *et al.*, 2009) and most importantly changes in fish and shrimp productivity.	The social implications of hydropower dams often include the resettlement of affected individuals and communities, psychological stress, the loss or decline of livelihoods, changes to lifestyles and traditions, impacts on fishing and agricultural activities as well as food security, impacts on access to and quality of water, and the range of environmentally adverse effects (Urban *et al.*, 2013).	Hydropower dam projects have often struggled in involving local stakeholders in participation, information and decision-making. Projects often fail to be transparent and accountable when it comes to informing and taking into account the social implications of hydropower dams for the most vulnerable stakeholders (Urban *et al.*, 2015).	The social implications of hydropower dams tend to be severe in low- and middle-income countries as these countries' social policies often rely upon informal security arrangements. In these cases stakeholders are dependent either upon their informal networks or upon the social arrangements provided by the hydropower dam projects. If these are lacklustre or not implemented, this often means severe problems for marginalised stakeholders (Urban *et al.*, 2015).

Figures provided by the International Energy Agency (IEA) indicate that in 2012 China's energy mix consisted of 88 per cent fossil fuels, of which coal accounted for 68 per cent, oil for 16 per cent and gas for 4 per cent (IEA, 2015). Only 17 per cent of China's total electricity supply came from hydropower (IEA 2015). There are often discrepancies between energy planning and realisation in relation to these energy statistics. In some cases, the Chinese data and the IEA's data differ to a certain degree. Recent research by Feng Wang and colleagues (2010) has shown that there are major barriers to China's low-carbon development:

- The share of renewable energy in the total energy mix is decreasing rather than increasing due to the rapid growth of fossil fuel capacity.
- The efficiency of renewable energy technology is low, as often the quality of the technology is not up to the latest standard and the efficiency of renewable energy technology is generally rather low.
- Many renewable energy technologies are not connected to the central grid, as this is costly and requires significant logistical resources. The installed renewable capacity is therefore often wasted. This could be interpreted as a market failure, particularly in cases where it is not financially viable for businesses to connect installed renewable energy technologies to the grid and where government authorities are reluctant to monitor implementation (Wang *et al.*, 2010: 1872).

This dependence upon fossil fuels should be seen as the backdrop to the will to invest in hydropower; a relatively clean energy source in a country plagued by pollution often caused by fossil fuels. Hydropower is a symbol for Chinese belief in large infrastructural projects and human engineering as a dual solution to China's issues of energy security and environmental problems. Under the new Chinese leadership of Xi Jinping and Li Keqiang, hydropower plays an important role. The 12th Five-Year Plan, covering the period from 2011 to 2015, foresees a domestic expansion of hydropower. The plan outlines the construction of major dams on key watersheds such as the Jinsha, Yalong and Dadu Rivers, and of new hydropower dam projects with a total installed capacity of 120 GW (China–Britain Business Council, 2011). With regard to Chinese dam projects overseas, there are currently 304 such dams, most of them in Southeast Asia (38 per cent) and Africa (27 per cent) (Urban and Nordensvärd, 2014). The largest hydropower dam-building company in China is the Sinohydro Corporation, a state-owned enterprise (SOE) that is subject to the rules and regulations of the State-Owned Asset Supervision and Administrative Commission (SASAC) (International Rivers, 2008a). Financial institutions like the ExIm Bank and Sinosure also have a significant interest in hydropower development and are involved in the majority of China's overseas investments (Heinrich Böll Stiftung/WWF/IISD, 2008). An estimated 60 per cent of Sinosure's medium- and long-term loan activity is for ExIm

Bank transactions (McDonald *et al.*, 2009). In addition to these large stakeholders, a number of non-state enterprises like Gezhouba participate in hydropower development and are involved in building and financing projects in over 30 countries in Asia and Africa, such as the Yeywa Dam in central Myanmar (McDonald *et al.*, 2009).

Chinese investments in hydropower have come at a large social price. In total, China's domestic dams are reported to have displaced 23 million people as well as significantly affected water availability and environmental quality (International Rivers, 2012). There has also been an issue with higher than expected construction costs due to bribery, excessive bureaucracy and corrupt local elites, such as has been suggested for the Chinese Three Gorges Dam project, where at least 349 local officials were allegedly found guilty of corruption (International Rivers, 2008a, 2008b; Hwang *et al.*, 2007; Middleton, 2008a). The project involved the flooding of 13 cities, 140 towns and 1,350 villages as well as numerous sites of cultural, historic and religious heritage. As a result, 1.3 million inhabitants were relocated, many to the cities (DTK, 2002; International Rivers, 2008a). It has been reported that many subsistence farmers and fishermen were relocated to cities or received tiny plots of barren land as compensation. As a result, many people were worse off after the resettlement due to the loss of livelihoods, increased unemployment, decreased income and insufficient land for subsistence farming. Furthermore, the resettlement process and the loss of cultural and social roots resulted in mental stress and an increase in depression among the resettled population (Hwang *et al.*, 2007; Urban *et al.*, 2013).

With regard to the excessive damming of China's rivers, it has been reported that a leading member of the Yangtze Valley Water Resources Protection Bureau believes there are too many dams on the rivers in China's southwest, which has had devastating environmental and social ramifications (Yan, 2012). At the same time, the Sichuan Geology and Mineral Bureau has published a study that claims that the Yangtze River might in the long term run dry as dam developers are planning to build an excessive number of large dams (Yan, 2012). The Chinese Institute of Public and Environmental Affairs argues that China is building too many dams on the Jinsha River, which is China's next large dam project with a capacity comparable to the Three Gorges Dam (Yan, 2012). After the experience of the Three Gorges Dam, some large hydropower projects have been stopped in China due to domestic and international opposition. A case in point is the planned hydropower development in the Tiger Leaping Gorge, which is one of China's most scenic and famous gorges. Construction of the dam was halted due to economic, cultural, social and environmental considerations. The project would have meant that large numbers of people would have lost their land and had to be resettled, a scenic tourist attraction would have disappeared, which would have had high economic impacts, and the cultural and natural heritage would have been lost. The Chinese government put these plans on hold and suggested that it had to learn from its mistakes made during the construction of other hydropower dams (International Rivers, 2008b; Urban *et al.*, 2013).

Considering these significant social and environmental impacts, it seems understandable that China is interested in investing in large hydropower projects beyond its own borders. China's rapid economic growth has also created three specific pressures that have forced it to engage more closely with a number of low- and middle-income countries:

1 Rapid growth is depleting already scarce domestic natural resources, including energy resources and minerals, and so part of China's 'Going Out Strategy' encourages overseas investment to access these resources (McNally *et al.*, 2009; Urban *et al.*, 2011).
2 As some sectors of the Chinese market become relatively saturated, the first generation of large SOEs liberalised under the post-1979 reforms need to internationalise and acquire new markets (Huang, 2008).
3 China's rapid technological advances – such as in energy technology – have made it possible to expand overseas.

These three drivers – resource access, new markets and technological advances – come together in the hydropower sector, where China is the pre-eminent global player in major dam projects, often with the support of Chinese state finance (Bosshard, 2009; Urban *et al.*, 2011). Power generation equipment is now China's second-largest export earner after electrical appliances (Bosshard, 2009).

China has become the most dominant international actor for international hydropower projects in recent years. While China has a long history of domestic dam-building, Chinese dam-builders are relatively new to the international dam industry and have rapidly increased their activity in recent years. China is today the world's largest hydropower developer. Chinese dam-builders often tend to invest in countries and regions where the World Bank, the Asian Development Bank and other multilateral organisations have stopped hydropower developments, such as in Myanmar and Sudan. This thereby offers opportunities for bringing infrastructure, resources and investments to poor and deprived countries. An example is the mining industry, which is reported to benefit from the access to electricity and water generated by the dams, such as the Nam Ngum 5 Dam in Laos and the Shweli 3 and Tarpein 1 Dams in Myanmar (International Rivers, 2011).

The number of Chinese actors involved in overseas hydropower development presents an increasingly complex picture. Chinese SOEs are becoming major players in extracting resources in riparian states along the Mekong. For instance, the ExIm Bank and other Chinese financial institutions, SOEs and private firms were involved in at least 93 major dam projects overseas in 2009 (McDonald *et al.*, 2009). As Amy McNally and colleagues highlight, this involvement cuts across many levels: "national-level power companies and their provincial subsidiaries, regional power grids, supra-regional yet sub-Ministerial basin (watershed) commissions, governmental units and legally-grounded citizen groups" (McNally *et al.*, 2009: S290).

Key actors include SOEs like the China National Water Resources and Hydropower Engineering Corporation (Sinohydro), the China Gezhouba Group Corporation (Gezhouba) and the China Southern Power Grid (CSG); financiers like the ExIm Bank, the China Export and Credit Insurance Corporation (Sinosure), the China Development Bank and the Bank of China; and regulators like the SASAC, the Ministry of Environmental Protection (MEP), the Ministry of Commerce (MOFCOM), the Ministry of Finance (MoF), the Ministry of Foreign Affairs (MoFA), the National Development and Reform Commission (NDRC) and others (Urban *et al.*, 2013).

In Southeast Asia as a whole, Chinese institutions are currently involved as contractors, developers, financiers and regulators in about 280 hydropower projects of large, medium and small size, including about 125 dams, in Brunei, Cambodia, Indonesia, Laos, Malaysia, Myanmar, the Philippines, Thailand and Vietnam. This is about 45 per cent of all Chinese overseas dams (International Rivers, 2011). Developments in Cambodia, Laos and Vietnam involve a small number of well-known key players such as Sinohydro and the ExIm Bank, whereas the developments in Myanmar involve a large number of smaller and less well-known players such as the Guangdong New Technology Import and Export Company, the Yunnan Machinery Export Import Company and China Hydropower Engineering Consulting, among others. As many of these hydropower deals are of a sensitive nature, some information is yet unknown, and many of these dams remain unresearched or under-researched. The amount of as yet unknown information makes it particularly difficult to assess the exact impacts of specific Chinese dam projects (Urban *et al.*, 2013).

Chinese-funded and -built hydropower in the Mekong region

The Mekong River represents an immensely valuable resource for China and East Asia. It is the eighth-largest river in the world, with a basin covering 800,000 square kilometres of mainland East Asia. Flowing through Yunnan province in southern China, it passes into Laos, Thailand, Cambodia and Vietnam and forms part of the border between Myanmar and China and between Myanmar and Laos (Goh, 2004). The Mekong is important for the livelihoods of the estimated 50 million people living in the basin, with the river supporting the largest freshwater fishery in the region, providing drinking water and a source of irrigation for rice cultivation and serving as a transportation route (Magee, 2006). In terms of power generation, Darrin Magee notes that "the Lancang's [Mekong's] annual hydropower generating capacity within Yunnan is estimated to be more than 100TWh" (Magee, 2006: 29). To put this figure into perspective, 80TWh per year would be enough to power the regions of Guangzhou, Shenzhen, Dongguan and Kunming combined. The significance of the Mekong River in generating hydropower is thus very high (Urban *et al.*, 2013). Table 3.2 indicates the hydropower potential in the Greater Mekong Subregion.

China draws on a history of dam construction that spans 45,000 large dams within its borders, built predominantly over the last 50 years (McDonald *et al.*, 2009). What makes dams in the Greater Mekong Subregion special and sets them apart from other world regions is that China can import electricity to power its booming urban areas. There is thus a direct link between creating hydroelectric capacity in Laos, Cambodia, Thailand, Vietnam and Myanmar and importing it as clean electricity to support the over-stretched domestic energy market. As a consequence, Laos has named itself the 'battery' of Asia and aims to export a large share of its electricity to neighbouring countries like China and Thailand, rather than using it domestically. In the coming years, Laos aims to increase its hydropower production capacity to 30,000MW from 670MW in 2008/9 (Jonsson, 2009: 202; Urban *et al.*, 2013).

While China has built dams on the Upper Mekong within its own borders for over a decade, the Lower Mekong has so far largely escaped hydropower development. However, as energy demand increases and the region's economy grows, there is an increasing demand for electricity in China as well as a hope of encouraging economic growth and investment in the countries of the Greater Mekong Subregion. The economic imperative behind the development of the Mekong River basin is captured in a statement by Khy Tainglim, Cambodia's Minister of Transportation: "Water is our oil [...] and we should use our water to export and get foreign currency to develop the country" (cited in Goh, 2004: 7).

These economic developments have come at an environmental and social price. The social implications of hydropower affect both individuals

Table 3.2 Estimated hydropower potential in the Greater Mekong Subregion

Country	*Hydropower resources (MW)*	*Technically and economically feasible resources (MW)*
China (Yunnan)	104,386	101,094
Myanmar	41,780	39,720
Laos	30,000	25,000
Vietnam	30,000	12,000
Thailand	10,620	9,130
Cambodia	10,000	8,600

Sources: China–Britain Business Council (2011).

and groups. According to William Adams, the social impacts of dams can be defined as "impacts on the lives of individual people or groups or categories of people, or forms of social organisation" (Adams, 2000: vi). The International Rivers database (2011) reveals a number of striking social ramifications:

- forced seizures of land from villagers at the Chibwe Dam in Myanmar;
- the displacement of at least ten villages and violence reported at the Myitsone Dam in Myanmar;
- allegations that local villagers were forced to work to construct buildings and roads for dam sites without remuneration at the Upper Paunglaung Dam and the Shweli 3 Dam in Myanmar;
- reports that assets of villagers such as land, livestock and natural resources were seized by the army at the Shweli 3 Dam;
- adverse effects on about 20,000 people at the Xeset 2 Dam in Laos;
- the resettlement of about 23,000 people at the Tuyen Quang Dam in Vietnam (Urban *et al.*, 2013).

In some cases, the Chinese dam sites are located in conflict areas that are already prone to violence, such as some border regions of Myanmar and Laos (International Rivers, 2011). Ideally, social impact assessments should be carried out to evaluate and mitigate the negative impacts on the local population; however, this is often missing or undertaken inadequately (Tilt *et al.*, 2009). The main debates revolve around whether positive outcomes in terms of economic development outweigh the costs, the unequal spatial distribution of the benefits and the impact on vulnerable groups like Indigenous people, women and reservoir evacuees (Adams, 2000). This results in a complex web of analysis, as impacts are both positive and negative, and even within communities, significant disparities occur along gender lines, which often go unseen in impact assessments of large dams. With regard to the immediate social impacts of large-scale dam construction, the main issues are:

- inappropriate resettlement;
- alterations in the lifestyle of people;
- the lowering of the standard of living;
- disregard for local people's rights, identity and culture;
- taking land and other natural resource tenure away from local people (Yu, 2003).

The trauma of resettlement is one of the biggest issues that communities have to grapple with. Loss of landholdings, insufficient compensation and cultural costs compounded with resettlement efforts that lower the standard of living of communities are among the immediate impacts (Adams, 2000). At the Upper Paunglaung Dam in Myanmar, for example, only about US$63 was reported to have been given as compensation for resettlement per household (International Rivers, 2011). The large hydropower development plans of China, however, have to be viewed in the context of their downstream impact on riparian states. Points of contention have been raised with regard to demands to protect fisheries, agriculture and shipping (McNally *et al.*, 2009). Furthermore, changes in aquatic life because of inundation and its subsequent impact on fisheries have not been sufficiently studied (Chang *et al.*, 2009). In the Lower Mekong, fisheries are a vital source of livelihood for an estimated 12 million rural households (Dore and Yu, 2004). Evelyn Goh (2004) adds that changes in the ecosystem such as higher water levels in the dry season and lower water levels in the wet season in Laos and Cambodia will significantly reduce the spawning ground for migratory fish. This in turn will affect productivity in wild-capture fisheries and the lifestyle of people in the region, for whom fish makes up 80 per cent of animal protein in their diet (Goh, 2004). One example where studies have been conducted to understand the potential impacts of dams on the Lancang is the Tonle Sap Great Lake in Cambodia, a vital ecological system supporting fisheries in the region and rice cultivation not only in Cambodia but also in southern Vietnam (Dore and Yu, 2004). Another case in point is the proposed construction of the Don Shaong hydropower project on the Lower Mekong River, which would block the main channel for fish migration (Jonsson, 2009; Urban *et al.*, 2013).

To address these concerns and criticisms from the international community, China passed an Environmental Impact Assessment (EIA) Law. According to this, companies proposing projects have to conduct an EIA prior to project construction, which then has to be approved by the appropriate environmental bureau (McDonald *et al.*, 2009). However, the potential of EIAs to address the above-mentioned issues is limited because there is "no requirement that EIAs be completed for policy or legislation [and] no provision for requiring that unavoidable impacts be mitigated by project developers" (McDonald *et al.*, 2009: S301). Transboundary Environment Assessment (EA) protocols and the UN Convention on the Law of the Non-Navigational Uses of International Watercourses have not yet had an impact on fostering cooperation or dialogue between riparian states to address the environmental ramifications of hydropower development (Dore and Yu, 2004). A major factor behind this is that not only is China the uppermost riparian state, but it is also the most politically powerful, has the fastest economic growth and remains relatively unfettered by the Mekong River Commission. Power relations between riparian states thus have a significant role to play in key decision-making (Urban *et al.*, 2013).

The case of the Kamchay Dam

The Kamchay Dam in Cambodia has a generating capacity of 193 MW and cost an estimated US$280 million (International Rivers, 2013), which was part of a US$600 million aid, trade and investment package from

China to Cambodia. The financiers were China's ExIm Bank, while the builders, developers and contractors were Sinohydro. Sinohydro started building the Kamchay Dam in Kampot province, southern Cambodia in 2006. The dam started operation in late 2011. After a restructuring of Sinohydro, the dam is now owned and operated by PowerChina Resources. The dam is subject to a Build, Operate, Transfer (BOT) contract, which means that the Cambodian government grants a concession to the dam-building firm for recovering its investment by allowing the dam-builder to own and operate the dam for 44 years and to sell electricity to the grid. This includes four years of construction time and 40 years of operation time. The dam is located in Bokor National Park, which obviously has significant environmental implications (Middleton, 2008b; Grimsditch, 2012).

While the Kamchay Dam is often treated as a case of adequate practice or a dam with minor impacts, as no resettlement of local people was required, the social impacts may appear less obvious but are nevertheless severe. This implies that one cannot generalise that Chinese dam-building is standardised or normalised. The analysis of the case study is presented in Table 3.3.

Table 3.3 Social implications of the Kamchay Dam in Cambodia

Substantive dimension	*Procedural dimension*	*Social policy dimension*
There has been no resettlement and therefore no displaced people. However,the livelihoods of people living downstream have been adversely affected, particularly poor bamboo collectors. The poorest may be forced to take up micro-credits or migrate to find work as unskilled labourers in factories in Phnom Penh or abroad to survive. Few social safeguards exist. There are no schemes for learning/self-development for the people adversely affected by the Kamchay Dam. There are no options for re-education or training.	Many villagers were not invited to participate in consultation processes and only became only aware of the dam once construction had started. There is also a range of unresolved complaints from the local population.	Ian Gough and Geof Wood (2006) define Cambodia as an insecurity regime with institutional arrangements that generate gross insecurity and block the emergence of stable informal mechanisms for the protection of individuals and communities. It is reliant upon powerful external players and dependent upon aid and remittances (Gough and Wood, 2006). Within this cluster they argue that Cambodia is a less effective informal security regime with poor levels of welfare coupled with low public commitments and moderate international inflows.

Substantive dimension

The Kamchay Dam has had a series of severe impacts on the local population. There are approximately 22,000 people living downstream in the rural area that is directly affected by the dam (NGO Forum on Cambodia, 2013). No resettlements have taken place to date as people did not live upstream due to the national park. The reservoir has been filled upstream in an area that was uninhabited. Only the better-off own land; most affected people are dependent upon informal labour to access food, income and housing.

As most employment is informal, different social groups have been affected to various extents by the hydropower dam. There are four groups of people who have been directly affected by the dam: bamboo collectors, fuelwood collectors, fruit sellers and durian growers.

The bamboo collectors are the largest group who have been adversely affected by the dam. They depend upon bamboo collection for their livelihoods, as they collect bamboo to make baskets. The baskets are sold on the local market in Kampot city. Most of the bamboo collectors do not have any other sources of income, many of them do not own any land or have any assets and most of them have very low literacy rates and can therefore not easily move on to more skilled jobs.

Before the dam was built, the villagers had an agreement with the Cambodian government that they would be allowed to collect bamboo in the 'multiple use zone' of Bokor National Park. They would usually start their trip in the morning, cycle up to the national park, collect the bamboo and be back home in the afternoon. However, the dam has flooded the bamboo forest area that the villagers used for bamboo collection. The dam has therefore meant a loss of or decline in livelihood for many bamboo collectors.

While the main bamboo collection area is flooded, the villagers have found another smaller bamboo forest site that is further away and now belongs to the land owned by Sinohydro/PowerChina Resources. Occasionally, the firm puts a ban on bamboo collection and closes off access to the area completely, sometimes for up to two weeks at a time (NGO Forum on Cambodia, 2013). This means that the villagers do not have any income for two weeks and instead have to borrow money from microcredit institutions. In the best-case scenario, Sinohydro/PowerChina Resources may open up access to the smaller bamboo site that is further afield. This means the villagers can collect bamboo; however, they either have to stay at the site overnight as it is far away or invest in a motorcycle to access the site, hence lowering productivity and cost-effectiveness. It is estimated that the affected local population have lost about 50 per cent of their daily income since the construction of the dam (Grimsditch, 2012). This sharp decline in income has posed a threat to livelihoods. Our fieldwork revealed that many of the villagers are therefore experiencing extreme financial hardship and adverse impacts on food security, and are considering options such as moving to Phnom Penh or even Thailand as migrant workers to secure their livelihoods.

The second group of people who have been affected by the dam are the fuelwood collectors. They face a similar situation to the bamboo collectors, although access to parts of the forest is currently banned by the Cambodian government for fuelwood collectors due to forest conservation efforts, and there is a ban on trucks and boats with engines on the reservoir.

The third group of people who have been negatively affected by the dam are the fruit sellers who are dependent upon income from tourism. Most of them buy fruits from local plantation owners and sell them at the Tuek Chhu riverside resort. The resort used to be popular with tourists, mainly from Phnom Penh, who came to swim in the river and enjoy the riverside. Since the dam has been built, the river is mostly dry in the dry season and to some extent also in the wet season. Water is only released sporadically by Sinohydro/PowerChina Resources according to a schedule that is unknown to both locals and authorities, as our interviews reveal. As the river is often dry, the affected locals reported that tourist numbers had declined by about 80 per cent in 2013, which is confirmed by other reports (NGO Forum on Cambodia, 2013). The sellers who depend upon tourism also mentioned that their incomes had declined by about 80 per cent since the construction of the dam, making it very difficult for them to support their families.

The fourth group of people who have been impacted by the dam are the durian growers and other plantation owners. Some of them received compensation from Sinohydro for land that was lost due to the construction of the dam (for example, land that was used for building roads). Rather than compensating for the lost land, the firm compensated the affected families for the lost trees. The villagers reported that a lost banana tree was compensated at US$10, a mango tree at US$30 and a durian tree at $100–500 depending upon size and age. The durian-growing families thought that this was a fair compensation payment. In fact, some of them said they benefit from the dam as it reduces flooding and thereby provides stable conditions for their flood-intolerant durian trees. Other groups of people who are directly affected by the dam are smallholder farmers who own land, such as rice fields, that has been acquired by Sinohydro to build construction infrastructure (mainly roads) and those who live directly under the power lines (ten families). Compensation was paid to villagers who lost their rice fields, but at only US$3 per square metre, which the villagers considered too low. No resettlement has taken place for those who live under the power lines and it is not yet clear how and when they will be compensated and relocated. Only a few people fish near the Kamchay Dam, most of them preferring to fish in the nearby sea. As no resettlement has taken place, housing and relocation are not an issue (see also Urban *et al.*, 2015).

Access to social infrastructure, local services and facilities has been mostly lacking in the case of the Kamchay Dam. Ironically, some villagers who live just next to the dam still do not have access to electricity, as it is being exported to Phnom Penh. Only about a third of the villages around the Kamchay Dam are reported to have access to electricity from the dam, and in these villages only about 15–20 per cent of households have access to electricity (NGO Forum on Cambodia, 2013). Nevertheless, the price of electricity has been reduced from 1,800 Riel per kWh to 920 Riel per kWh; however, this is higher than the initially mentioned 500–600 Riel per kWh that was promised by Sinohydro (Middleton, 2008b). Even though electricity has become more affordable, many people do not have the financial means to connect to the grid as it requires a connection fee of US$160 per household, as the villagers report. Basic water access is provided, as it was before the dam's construction. Some people who are better off, such as the durian plantation owners, live in adequate housing, while poorer local people, such as some of the bamboo collectors, live in impoverished and crowded conditions in ramshackle houses. As their livelihoods are negatively affected by the dam, it is unlikely that their access to adequate housing, sanitation, electricity and water will improve any time soon (see also Urban *et al.*, 2015).

There are no schemes for learning/self-development for the people adversely affected by the Kamchay Dam, and neither are there any options for re-education or training. The only two options that affected individuals have are either to migrate to find work as unskilled labourers in factories in Phnom Penh or to take up micro-credits to be able to pay for food and other costs. In general, far fewer social safeguards exist with regard to the Kamchay Dam compared to the Bui Dam in Ghana and neither Sinohydro/PowerChina Resources nor the Cambodian government have made serious efforts to change this.

The overall impact on community/social cohesion was not discussed in detail by the affected people; however, considering migration as a last resort to ensure survival will have a devastating impact on social cohesion and the structure of communities.

Environmental bads are shared unequally in the affected communities. Those who are already worse off, such as the bamboo collectors, are disproportionately negatively affected by the dam, while those who are better off, such as the durian plantation owners, benefit modestly from the dam. Access to natural resources such as timber and non-timber forest resources is restricted for the local population as parts of Bokor National Park now belong to Sinohydro/PowerChina Resources and have been flooded. In addition, Sinohydro/PowerChina Resources receives the revenue from generating electricity from the dam for 44 years while the Cambodian government and Cambodian firms lose out on this revenue.

Procedural dimension

The governance and management process for the Kamchay Dam project has been markedly different from that of the Bui Dam. The Kamchay Dam contract between Sinohydro/PowerChina Resources and the Cambodian government is a BOT contract. Sinohydro/PowerChina Resources will transfer the ownership of the dam to the Cambodian government in 2050, after 44 years (four years of construction time and 40 years of operation time). In the 1990s, before Sinohydro won the contract, the Canadian International Development Agency (CIDA) was in negotiations with the Cambodian government about building the dam, but withdrew because of pressure from local and international NGOs due to the social and environmental impacts of the dam (Grimsditch, 2012).

The government authorities that are responsible for dams in Cambodia are the Ministry of Industry, Mines and Energy (MIME), the Ministry of Water Resources and Meteorology (MOWRAM) and the Ministry of Environment (MoE). All BOT projects such as the Kamchay Dam have to be approved by the Council for the Development of Cambodia (CDC). There is a Law on Water Resource Management and all hydropower projects require a water use license from MOWRAM. In addition, Cambodia has a Law on Environmental Protection and Natural Resource Management, a Forestry Law and a complex yet underdeveloped Land Law that have to be complied with (Grimsditch, 2012).

Accountability is generally weak for failings related to the dam. There have been serious shortcomings with regard to the Environmental and Social Impacts Assessment (ESIA). By Cambodian law, development projects such as dams are required to have an EIA in place and approved before the construction process begins. Cambodian law also prescribes that the EIA process should be transparent, the decision-making should be accountable and a wide consultation process should involve affected local communities and civil society organisations (Middleton, 2008b; NGO Forum on Cambodia, 2013). However, in the case of the Kamchay Dam, the EIA process was seriously flawed. The EIA process started late and the EIA approval was in fact granted seven months after the inauguration of the dam (International Rivers, 2013; NGO Forum on Cambodia, 2013). It has been reported by several experts that the content of the ESIA is of poor quality as, for example, it does not assess the impacts of the dam on species and habitats but only lists what species occur in the national park. Overall, the implementation of environmental and social safeguards is minimal and not in line with Cambodian legislation, as the interviews reveal. In addition, the Environmental Management Plan (EMP), which aims to implement mitigation measures to reduce the negative effects of the dam, was not in place until the late stages of the dam construction. It is also reported that Sinohydro/PowerChina Resources refuses to implement any mitigation measures, as confirmed by our interviews and other reports (NGO Forum on Cambodia, 2013). The firm is said to have set aside an as-yet-untouched budget of US$5 million for implementing mitigation measures; however, even high-ranking government officials tend to criticise Sinohydro/PowerChina Resources for its inaction. In addition, many locals were not consulted prior to the dam's construction and public participation therefore existed only on paper rather than in practice. The private Cambodian company that was contracted to conduct the ESIA, SAWAC, reportedly spent about 80 per cent of its ESIA budget on avoiding red tape to ease the paperwork for the ESIA approval. Given these figures, unofficial reports of bribery have been made (see also Urban *et al.*, 2015).

The communication and decision-making process between Sinohydro/PowerChina Resources and the Cambodian authorities seems to have been rather opaque and very hierarchical, taking place at the national level in Phnom Penh rather than at the level of the provincial government agencies and Sinohydro's/PowerChina Resources' base at the dam site.

The environmental and social monitoring process has been to the responsibility of the Cambodian state. The main legal framework for the EIA is the Sub-decree on EIA passed by the MoE in 1999. The EIA has to be approved by several ministries, including those mentioned above. The MoE is primarily responsible for organising the undertaking of the EIA, reviewing the report and monitoring compliance with environmental legislation (Grimsditch, 2012).

There is concern about the EMP as no mitigation measures have been undertaken so far. It is also unclear who is responsible for implementing the EMP. The MoE and the provincial Department for the Environment (DfE) are supposed to monitor the implementation of the EMP and are waiting for Sinohydro's/PowerChina Resources' action. The dam operator, on the other hand, is expecting the MoE and the provincial DfE to act first, and as a result no action has been taken so far.

Access to participation and decision-making for the stakeholders and proactive stakeholder communication and consultation have been neglected throughout the project. As part of the Cambodian EIA legislation, every EIA needs to be approved before the construction starts and consultation with all stakeholders is required. This clearly did not happen for the Kamchay Dam as the EIA approval came after the dam had started operation and many locals who are directly affected by the dam have not been consulted. Other social issues include a lack of consultation for the dam construction and the ESIA. The consultation process before the dam's construction was patchy and ad hoc with little local participation, as our fieldwork finds and other reports confirm (International Rivers, 2013). Many villagers were not invited to consultation processes and only became aware of the dam once construction had started. There is also a range of unresolved complaints from the local population, for example relating to the closure of the bamboo forest area and the ban on boats with engines. Several demonstrations took place in the provincial capital Kampot, with several hundred local people demonstrating at government departments. In addition, several petitions have been signed against the firm's management of the dam, most importantly petitions against the closure of the bamboo forest area.

The communication and decision-making process between the local population and Sinohydro/PowerChina Resources is even more complicated. As mentioned before, the local villagers have complained in various forms (petitions, mass demonstrations, individual complaints) against Sinohydro/PowerChina Resources. Nevertheless, they have had to follow a strict hierarchical procedure, addressing first the village chief, then the commune authority, then the district authority, then the provincial authority, and from then on the complaints are said to be taken to the appropriate ministries in Phnom Penh (mainly MIME) which then establish communication with the dam operator. This is despite Sinohydro's/PowerChina Resources' offices being based at the dam site, in very close proximity to the affected villages (see also Urban *et al.*, 2015).

A major problem occurred in September 2015, when heavy rains meant that the dam exceeded its maximum capacity. Sinohydro/PowerChina Resources released the floodgates and literally flooded the communities adjacent to the dam. While no one is reported to have died, the homes of several thousand people were flooded (with estimates ranging from 500 to 3,600 affected families), as well as their rice fields and agricultural land. The true extent of the flooding may have been even worse as Kampot city, several kilometres away from the dam, was also flooded as a result of the dam's floodgates being opened (Chakrya and Chandara, 2015; Channyda, 2015; Laurenson, 2015; Odom, 2015; Odom and Blomberg, 2015; Yang, 2015). The poorest were hit the hardest, as their homes are the most flimsy, precarious and vulnerable to floods. They are already struggling to provide income, food and medicine for their families, yet were particularly exposed to the flooding. Media reports mentioned that in some places the flood level was up to two metres high. It has been reported that the Chinese dam operator notified the Cambodian authorities one day in advance of the flooding; however, the Cambodian authorities only notified the local communities one hour before the flooding, or even later (Chakrya and Chandara, 2015; Channyda, 2015; Laurenson, 2015; Odom, 2015; Odom and Blomberg, 2015; Yang, 2015). This incident was a tragic failure to evacuate the people living next to the Kamchay Dam and a sign that neither the Cambodian authorities nor the Chinese dam operator feel responsible for the impact that the dam has on local people. Meanwhile, there were rumours among local people that the floodgates were opened on purpose (VOA, 2015). Ironically, the governor of Kampot province argues that the Kamchay Dam serves the purpose of flood control. This can easily be challenged given the recent flooding caused by suddenly opening the floodgates. This incidence clearly demonstrates the social injustice that local people face due to the nearby dam and the mismanagement of the dam operator and the local government, as well as the urgent need to re-evaluate the suitability of further dams planned in Cambodia.

Social implications, challenges and solutions

This chapter has explored the social implications of constructing hydropower dams such as the relocation of people and the loss of livelihoods of people affected by the dam sites. The chapter has also explored the conflict that arises at the sites of the dams and how hydropower dams could be built in the future with far fewer social ramifications. There has been a high impact on livelihoods at the Kamchay Dam as most locals do not have alternative livelihood strategies. One must distinguish between the different roles that Chinese institutions have in the process of dam-building. There are financiers/investors, developers, builders and contractors, component suppliers and dam operators (Urban *et al.*, 2013):

- *Financiers/investors* include the ExIm Bank, which as a funder of large dams could be considered equal to the World Bank, which also funds large dams. Financiers/investors such as the ExIm Bank may not have much power over social issues such as resettlements and compensation for affected people, but they could refrain from investing in morally and socially irresponsible projects. They should also be bound by the Equator Principles for investors.
- *Developers* may be influenced by the host government, for example with regard to site selection. However, they may be involved in initiating surveys and planning the dam, including its design and construction, which often includes major decisions about the size of the dam, its generating capacity, the size of the reservoir that will flood the area, and so forth. This may also include contributing to major decisions about whether individual people, villages, towns, agricultural land and cultural sites will be affected by the dam-building and the flooding of the area, how many people will be affected and whether they will be resettled. Depending upon the type of contract, and particularly for BOT contracts, this may also include decisions about compensation payments, including how much money should be made available. Developers may also be involved in requesting an EIA or an ESIA so that the environmental and social impacts of the dams are evaluated and mitigation measures are developed (such as resettlements and compensation, rescue missions for animals before the reservoir flooding, planting new trees after the flooding, and so forth). Developers should therefore be well aware of the full extent of the dam-building process and its impacts and able to make informed decisions about its sustainability and whether or not a project should go ahead.
- The *builders* of the dam or the *contractors* who either build the dam entirely or supply components (such as the turbines) are in charge of

the engineering and the actual construction and implementation of the dam, sometimes including connecting it to the grid. While they may not be aware of the full extent of the socially and environmentally adverse effects of the dam, they are usually sufficiently informed to decide whether or not the project is morally and socially defensible.

- *Dam operators* are those who are in charge of managing the dam on a daily basis. Like the builders and contractors, they may not be aware of the full extent of the socially and environmentally adverse effects of the dam, but they are usually sufficiently informed to decide whether or not the project is morally and socially defensible.

Sometimes these roles are divided between different actors; sometimes one actor has multiple roles, such as being developer, builder and dam operator all in one, as was the case with Sinohydro's/PowerChina Resources' role for the Kamchay Dam (see also Urban *et al.*, 2015). Table 3.4 sums up some of the social consequences of the Chinese-built and Chinese-funded Kamchay Dam.

The problems of relocation, loss of livelihoods and environmental degradation have made hydropower one of the most controversial low-carbon

Table 3.4 Social implications, challenges and solutions

Social implications	*Challenges*	*Solutions*
The problems of relocation, loss of livelihoods and environmental degradation have made hydropower one of the most controversial low-carbon technologies. In low- and middle-income countries, these problems hit the poorest harder as they rely upon informal security arrangements, informal employment and their direct environment.	Hydropower dams such as the Kamchay Dam in Cambodia highlight two main social challenges. On one side, the issues of undermining the local stakeholders highlight the lack of overall formal social policies to lighten the burden on the poorest. Those reliant upon informal welfare and informal employment are hit hardest by a hydropower dam. On the other side, there is an assumption that international hydropower projects will be able to mitigate some of the negative social implications, which in reality is often not the case.	There is a stark contrast between formal national and regional social policies that should mitigate social implications and the practical social policy measures that are offered through hydropower dam projects. A superficial solution would be to improve dam-building contracts to include more extensive social policy/ social protection, but in the end it will be a matter of global regulation for investment and stronger national policies to include social policy and social and environmental sustainability at a minimum level for all international hydropower dam projects.

technologies. In low- and middle-income countries, these problems hit the poorest harder as they rely upon informal security arrangements, informal employment and their direct environment.

Hydropower dams such as the Kamchay Dam in Cambodia highlight two main social challenges. On one side, the issues of undermining the local stakeholders highlight the lack of overall formal social policies to lighten the burden on the poorest. Those reliant upon informal welfare and informal employment are hit hardest by a hydropower dam. On the other side, there is an assumption that international hydropower projects will be able to mitigate some of the negative social implications, which in reality is often not the case.

Wood and Gough (2006) define Cambodia as an insecurity regime with institutional arrangements that generate gross insecurity and block the emergence of stable informal mechanisms for the protection of individuals and communities. It is reliant upon powerful external players and dependent upon aid and remittances (Wood and Gough, 2006). Within this cluster, they argue that Cambodia is a less effective informal security regime with poor levels of welfare coupled with low public commitments and moderate international inflows (Wood and Gough, 2006).

A superficial solution would be to improve dam-building contracts to include more extensive social policy/social protection, but in the end it will be a matter of global regulation for investment to include social policy and social and environmental sustainability at a minimal level for all international hydropower dam projects. The Kamchay Dam is an example of a BOT contract, which means that the Cambodian government grants a concession to Sinohydro/PowerChina Resources for recovering its investment by allowing the dam-builder to own and operate the dam for 44 years and to sell electricity to the grid. After the 44 years of ownership by Sinohydro/PowerChina Resources, the dam ownership will be transferred to the Cambodian government, albeit after the peaking of the economic lifetime of the dam. Technically, this makes the dam-builder Sinohydro/PowerChina Resources in charge of the resettlement process (which was not needed at the Kamchay Dam), of compensating the affected local people, of implementing mitigation measures to reduce the adverse environmental and social impacts of the dam and of managing the dam and its impacts on a daily basis. In reality, however, Sinohydro/PowerChina Resources operates and manages the dam, but leaves dealing with the social and environmental impacts of the dam largely to the local authorities. Some Cambodian authorities in turn expect Sinohydro/PowerChina Resources to act, for example by referring to the outstanding payments for mitigation measures such as the provincial DoE. Other Cambodian authorities have dealt with compensation payments but are ignoring further local discontent.

While a BOT contract technically makes the contractor, builder and operator (in this case Sinohydro/PowerChina Resources) in charge of the dam and its social and environmental impacts, in the case of the Kanchay

Dam these cumbersome tasks seem to have been left to the Cambodian authorities to deal with. Other projects have had different solutions. For example, the Bui Dam in Ghana is an Engineering, Procurement, Construction (EPC) contract, which is a turnkey contract in which the Chinese dam-builder Sinohydro is the contractor who builds the dam and then transfers the ownership and daily management of the dam to Ghana's government. This makes Ghana's government in charge of the resettlement process, of compensating the affected local people, of implementing mitigation measures to reduce the adverse environmental and social impacts of the dam and of managing the dam and its impacts on a daily basis. Following the transfer of ownership, Ghana's government established the Bui Power Authority to manage the dam and its impacts.

While the development of water resources for long-term economic development is also a motivation for the countries of Southeast Asia to construct dams, it is important to view this in the context of the existing power disparities between China and countries along the Lower Mekong. In Cambodia, for example, China is a donor and an investor in infrastructure projects like roads and bridges as well as a trading partner (Goh, 2004). Dams are often part of the aid packages that China provides to foreign governments, a method that was adopted by other East Asian countries like Japan or Korea in the past. This creates complications for affected countries in expressing concerns over the ecological and social impacts of the construction of dams along the Mekong River. Despite these adverse effects, hydropower can be environmentally beneficial as it mitigates high amounts of greenhouse gas emissions and is thereby a climate-friendly form of energy. It can replace significant amounts of coal and thereby contribute to a better climate, reduced air pollution and reduced fossil fuel resource depletion. This could contribute to sustainable development in China, whereas hydropower's sustainability in the electricity-exporting countries could be questioned due to the significant environmental and social impacts of dams and the unequal distribution of benefits. On a different note, Chinese companies are simultaneously conducting feasibility studies for dams and serving as financiers, builders and regulators of hydropower projects, resulting in a blurring of lines between these roles. The sheer multiplicity of public and private actors involved in the hydropower industry also raises transparency and accountability issues. While the Chinese government has provided guidelines to protect the interests of local employees and maintain environmental standards, often these guidelines are only partly effective as Chinese enterprises operate at a distance from the government, particularly when these norms are in conflict with private interests (Bosshard, 2009). Often, the implementation of these social and environmental guidelines depends as much upon the host country as the Chinese institutions involved. However, over the past few years, China's environmental laws have been strengthened with the introduction of improved EIAs, which call for public participation and the approval of the Ministry of Environmental Protection (International Rivers, 2008a).

In addition, laws have also been instituted on the displacement of people, thus including provisions for resettlement and for compensation for loss of livelihood at a level similar to or greater than the original livelihoods of the displaced people. While this law currently pertains to dams within China's borders, it can be used as a guideline for overseas projects as well. With respect to overseas projects, Nine Principles Governing the Activities of Foreign Investment Firms were issued in 2006 by China's State Council, which called for mutual respect, complying with local laws and protecting environmental resources (Middleton, 2008a). Sinohydro, China's largest dam-building SOE, has pledged to develop a consistent environmental policy across its group, while China Southern Power Grid, another major hydropower company, has only a very general policy document. The question that persists is whether the new national laws will be applied to overseas projects in the future and will thereby ease some of the ramifications for people dependent upon the Mekong and its environment. China has set up ambitious social and environmental regulations that have often evaporated in practice.

The no-strings-attached policy might be little more than conceding problems in implementing environmental and social regulations and codes. The economic and public costs of hydropower dams in environmentally and socially sensitive areas makes Western corporations sometimes wary of these investment opportunities. China has started to rethink its no-strings-attached policy. In January 2007, Cheng Siwei, Vice-Chairman of the Standing Committee of the People's Congress, warned that "irresponsible practices" had prevented Chinese companies from expanding their businesses abroad. He predicted: "Even in developing countries, foreign companies that turn a blind eye to their social responsibilities will be kicked out of the market" (McDonald *et al.*, 2009: S302). Faced with this external pressure, Sinohydro developed a manual on health, safety and the environment for overseas investments in 2013, which was meant to provide guidelines on how to deal with some of the negative impacts of the dam-building industry. Other dam-builders may have other guidelines; however, many do not have any standardised guidelines at all.

Social sustainability and social policy varies from dam to dam as Chinese dam-builders lack an overall standardised code of conduct regarding social impacts. Each individual hydropower project is dependent upon the local conditions of governance, the type of contract and the type of role that the Chinese dam-builders play. As such, the approach to social sustainability is not necessarily exported from China to other countries but depends upon a range of factors that determine the power relations between Chinese dam-builders and the host countries. There are asymmetric power relationships between wealthy Chinese dam-builders and poorer host countries (such as Cambodia), which may give dam-builders a large degree of freedom to manoeuvre without much accountability and host countries little opportunity to negotiate or make demands. However, a change in attitude can be observed as China's investment operations in large dams in poorer countries are being closely monitored internationally by NGOs, academics, industry and governments, which can influence the wider reputation of both the dam-builders and China. Some improvements can be seen, such as Sinohydro's recently established manual on health, safety and the environment for overseas dam projects.

Another of the lessons that we learn from the Kamchay Dam is that that the national context is one of the most decisive factors that influences how international dam-building firms deal with the social and environmental impacts of large dams (Hensengerth, 2013), such as compensation, EIAs, complying with international standards and national laws, and dam emergencies such as the recent flooding.

There is a need to set up a global benchmarking system and a legally binding global code of conduct for social sustainability for international hydropower projects. Part of this is already in existence – for example, International Rivers' benchmarking report for large Chinese dam-builders (International Rivers, 2015). This could be extended and applied for dam-building firms worldwide. Standards formulated by the World Bank/IFC, the IHA's Hydropower Sustainability Assessment Protocol, the recommendations of the World Commission on Dams and the Equator Principles for financial institutions are already in place. The problem is that these standards and codes of conduct are not legally binding at present and depend upon the voluntary actions of dam-builders. In addition to making these legally binding, it would be useful for future dam contracts to be regulated by an international governing body to prevent severe social implications for the local population. This applies not only to Chinese dam-builders, but also to other dam-builders from any other country or organisation.

At the moment these measures are not in place and dam-builders and some complicit national governments are likely to object to these attempts. As long as this happens, the international community including NGOs, activists, experts and academics will serve as observers and judges of dam-building firms and their environmental and social impacts.

Acknowledgements

Part of this chapter is taken and adapted from Urban and colleagues (2015), with kind permission from John Wiley & Sons. Another part of the chapter is taken and adapted from Urban and colleagues (2013), with kind permission from Springer Science & Business Media.

References

Adams, W. (2000), *The Social Impact of Large Dams: Equity and Distribution Issues*, WCD Thematic Review Social Issues I.1, prepared for the World Commission on Dams, Cape Town: WCD, available at: www.esocialsciences.org/Download/repecDownload.aspx?fname=Document1115200600.5004389.pdf&fcategory=Articles&AId=513&fref=repec (accessed 20 March 2016).

Ampiah, K. and Naidu, S. (2008), "The Sino-African relationship", in K. Ampiah and S. Naidu (eds), *Crouching Tiger, Hidden Dragon: Africa and China*, Scottsville: University of KwaZulu-Natal Press, pp. 329–39.

Bosshard, P. (2009), "China dams the world", *World Policy Journal*, 26(4), 43–51.

Chakrya, K. S. and Chandara, S. (2015), "Floods wreak havoc", *Phnom Penh Post*, 18 September, available at: http://m.phnompenhpost.com/national/floods-wreak-havoc (accessed 20 March 2016).

Chang, X. L., Liu, X. and Zhou, W. (2009), "Hydropower in China at present and its further development", *Energy*, 35(11), 4400–6.

Channyda, C. (2015), "Kampot households plagued by flooding", *Phnom Penh Post*, 17 September, available at: www.phnompenhpost.com/national/kampot-households-plagued-flooding (accessed 20 March 2016).

China–Britain Business Council (2011), *China's Twelfth Five-Year Plan (2011–2015)*, available at: www.britishchamber.cn/content/chinas-twelfth-five-year-plan-2011-2015-full-english-version (accessed 27 May 2016).

Dore, J., and Yu, X. (2004), "Yunnan hydropower expansion: update on China's energy industry reforms and the Nu, Lancang and Jinsha hydropower dams", Working Paper from Chiang Mai University's Unit for Social and Environmental Research and Green Watershed, available at: www.sea-user.org/download_pubdoc.php?doc=2586 (accessed 20 March 2016).

DTK (German Dam Committee) (2002), "The Three Gorges project at the Yangtze: provisions and reality", available at: http://talsperrenkomitee.de/das_three_gorges_project_am_yangtze/das_three_gorges_project.htm (accessed 20 March 2016).

Goh, E. (2004), "China in the Mekong River basin: the regional security implications of resource development on the Lancang Jiang", RSIS Working Paper No. 69, Singapore: Nanyang Technological University, available at: http://dr.ntu.edu.sg/bitstream/handle/10220/4469/RSIS-WORKPAPER_73.pdf?sequence=1 (accessed 20 March 2016).

Grimsditch, M. (2012), *China's Investments in Hydropower in the Mekong Region: The Kamchay Hydropower Dam, Kampot, Cambodia*, available at: www.bicusa.org/wp-content/uploads/2013/02/Case+Study+-+China+Investments+in+Cambodia+FINAL+2.pdf (accessed 20 March 2016).

Hall, A. (1994), "Grassroots action for resettlement planning: Brazil and beyond", *World Development*, 22(12), 1793–809.

Hall, A. (2007), "Social policies at the World Bank: paradigms and challenges", *Global Social Policy*, 7(2), 151–75.

Heinrich Böll Stiftung/WWF/IISD (International Institute for Sustainable Development) (2008), "Rethinking investments in natural resources: China's emerging role in the Mekong region", policy brief, Heinrich Böll Stiftung/WWF/IISD, available at: www.iisd.org/pdf/2008/trade_chinapolicybrief.pdf (accessed 27 May 2016).

Hensengerth, O. (2013), "Chinese hydropower companies and environmental norms in countries of the Global South: the involvement of Sinohydro in Ghana's Bui Dam", *Environment, Development and Sustainability*, 15(2), 285–300.

Huang, Y. (2008), *Capitalism with Chinese Characteristics: Entrepreneurship and the State*, Cambridge: Cambridge University Press.

Hwang, S. S., Xi, J., Cao, Y., Feng, X. and Qiao, X. (2007), "Anticipation of migration and psychological stress and the Three Gorges Dam project, China", *Social Science and Medicine*, 65(5), 1012–24.

IEA (International Energy Agency) (2015), "Energy statistics: China data", available at: www.iea.org/statistics/ (accessed 20 March 2016).

International Rivers (2008a), "The Three Gorges Dam: the cost of power", fact sheet, Berkeley, CA: International Rivers, available at: www.internationalrivers.org/resources/three-gorges-dam-the-cost-of-power-2651 (accessed 20 March 2016).

International Rivers (2008b), "Delegate calls for Tiger Leaping Gorge rethink", Berkeley, CA: International Rivers, available at: www.internationalrivers.org/files/attached-files/delegate_calls_for_tiger_leaping_gorge_rethink.pdf (accessed 20 March 2016).

International Rivers (2011), "Database of China's dam building projects in Southeast Asia", available at: www.internationalrivers.org/node/3110 (accessed 20 March 2016).

International Rivers (2012), "China's major rivers", available at: www.internationalrivers.org/campaigns/china-s-major-rivers (accessed 27 May 2016).

International Rivers (2013), "Cambodia", available at: www.internationalrivers.org/campaigns/cambodia (accessed 20 March 2016).

International Rivers (2015), *Benchmarking the Policies and Practices of International Hydropower Companies: Stage 1 – Environmental and Social Policies and Practices of Chinese Overseas Hydropower Companies*, Berkeley, CA: International Rivers, available at: www.internationalrivers.org/files/attached-files/benchmarking_report_english_part_a.pdf (accessed 20 March 2016).

Jonsson, K. (2009), Laos in 2008: hydropower and flooding (or business as usual)", *Asian Survey*, 49(1), 200–5.

Laurenson, J. (2015), "Thousands affected by Kampot flood", *Khmer Times*, 17 September, available at: www.khmertimeskh.com/news/15864/thousands-affected-by-kampot-floods/ (accessed 20 March 2016).

Macdonald, E. (2001), "Playing by the rules: the World Bank's failure to adhere to policy in the funding of large-scale hydropower projects", *Environmental Law*, 31(4), 1011–49.

McDonald, K., Bosshard, P. and Brewer, N. (2009), "Exporting dams: China's hydropower industry goes global", *Journal of Environmental Management*, 90, S294–S302.

McNally, A., Magee, D. and Wolf, A. T. (2009), "Hydropower and sustainability: resilience and vulnerability in China's powersheds", *Journal of Environmental Management*, 90, S286–93.

Magee, D. (2006), "Powershed politics: Yunnan hydropower under Great Western development", *China Quarterly*, 185, 23–41.

Middleton, C. (2008a), "The sleeping dragon awakes: China's growing role in the business and politics of hydropower development in the Mekong Region", Berkeley, CA: International Rivers, available at: www.internationalrivers.org/files/

attached-files/sleepingdragonawakes_watershed_nov08_text_only.pdf (accessed 20 March 2016).

Middleton, C. (2008b), *Cambodia's Hydropower Development and China's Involvement*, Phnom Penh: Rivers Coalition Cambodia.

Mirumachi, N. and Torriti, J. (2012), "The use of public participation and economic appraisal for public involvement in large-scale hydropower projects: case study of the Nam Theun 2 hydropower project", *Energy Policy*, 47, 125–32.

Mohan, G. and Power, M. (2008), "New African choices? The politics of Chinese engagement", *Review of African Political Economy*, 35(115), 23–42.

NGO Forum on Cambodia (2013), *The Kamchay Hydropower Dam: An Assessment of the Dam's Impacts on Local Communities and the Environment*, Phnom Penh: NGO Forum on Cambodia.

Nordensvärd, J. and Urban, F. (2015), "Social innovation and Chinese overseas hydropower dams: the nexus of national social policy and corporate social responsibility", *Sustainable Development*, 23(7–8), 245–56.

Odom, S. (2015), "Hundreds flee homes as dam gates opened", *Cambodia Daily*, 17 September, available at: www.cambodiadaily.com/news/hundreds-flee-homes-after-dam-gates-opened-94476/ (accessed 20 March 2016).

Odom, S. and Blomberg, M. (2015), "Kampot, B'bang still flooded as rains continue", *Cambodia Daily*, 18 September, available at: www.cambodiadaily.com/news/kampot-bbang-still-flooded-as-rains-continue-94679/ (accessed 20 March 2016).

Tilt, B., Braun, Y. and He, D. (2009), "Social impacts of large dam projects: a comparison of international cases: studies and implications for best practice", *Journal of Environmental Management*, 90, S249–57.

Urban, F. and Nordensvärd, J. (2014), "China dams the world: the environmental and social impacts of Chinese dams", *E-International Relations*, 30 January, available at: www.e-ir.info/2014/01/30/china-dams-the-world-the-environmental-and-social-impacts-of-chinese-dams/ (accessed 20 March 2016).

Urban, F., Mohan, G. and Zhang, Y. (2011), "The understanding and practice of development in China and the European Union", IDS Working Paper 372, Brighton: Institute of Development Studies.

Urban, F., Nordensvärd, J., Khatri, D. and Wang, Y. (2013), "An analysis of China's investment in the hydropower sector in the Greater Mekong Subregion", *Environment, Development and Sustainability*, 15(2), 301–24.

Urban, F., Nordensvärd, J., Siciliano, G. and Li, B. (2015), "Chinese overseas hydropower dams and social sustainability: the Bui Dam in Ghana and the Kamchay Dam in Cambodia", *Asia & the Pacific Policy Studies*, 2(3), 573–89.

VOA (Voice of America Khmer) (2015), "Floods in Kampot not created on purpose, governor says", 21 September, available at: www.voacambodia.com/content/floods-in-kampot-not-created-on-purpose-governor-says/2972417.html (accessed 20 March 2016).

Wang, F., Yin, H. and Li, S. (2010), "China's renewable energy policy: commitments and challenges", *Energy Policy*, 38(4), 1872–8.

Watson, J. and Wang, T. (2007), "Who owns China's carbon emissions?", Tyndall Centre Briefing Note No. 23, Norwich: Tyndall Centre.

WCD (World Commission on Dams) (2000), *Dams and Development: A New Framework for Decision-Making*, London: Earthscan.

Wood, G. and Gough, I. (2006), "A comparative welfare regime approach to global social policy", *World Development*, 34(10), 1696–712.

Yan, K. (2012), "New wave of Three Gorges-sized dams raise old fears", *International Rivers*, 7 June, available at: www.internationalrivers.org/blogs/246/new-wave-of-three-gorges-sized-dams-raise-old-fears (accessed 20 March 2016).

Yang, C. (2015), "Flooding from dam washes out homes, rice fields in Southern Cambodia", *Radio Free Asia*, 16 September, available at: www.rfa.org/english/news/cambodia/flooding-from-dam-washes-out-homes-rice-fields-in-southern-cambodia-09162015162039.html (accessed 20 March 2016).

Yu, X. (2003), "Regional cooperation and energy development in the Greater Mekong Subregion", *Energy Policy*, 31(12), 1221–34.

4 The high costs of wind energy in Germany

Social challenges and possible solutions

Johan Nordensvärd with Frauke Urban

Wind energy in Germany

Wind energy, along with hydropower and solar energy, is considered to be an important technology in spearheading an energy transition (*Energiewende*) to sustainable and low-carbon energy resources in Germany. Germany is the EU's largest wind energy market, the world's third-largest market for wind energy (after China and the US) and a global forerunner in wind energy innovation (BWE, 2012; GWEC, 2014). Germany had an installed wind energy capacity of more than 34 GW by the end of 2013. This accounted for about 30 per cent of the European installed wind capacity in 2013 (GWEC, 2014; IEA, 2014). Germany's installed capacity and market has been growing continuously since the mid-1990s (BWE, 2012; IEA, 2014). Germany has both considerable onshore wind capacity and a rapidly growing offshore capacity. The German government has targets in place for a share of 35 per cent renewable energy of the total electricity mix by 2020, 50 per cent by 2030 and 80 per cent by 2050, with respect to which wind plays an important role (BMU, 2012; BMU, 2011). The market shares of wind firms in Germany are as follows:

- Enercon – about 60 per cent;
- Vestas – 20 per cent;
- REpower – 10 per cent;
- Nordex – 4 per cent;
- Bard – 2 per cent;
- others including e.n.o., Vensys, Siemens GE Electric and AREVA – remaining 4 per cent (Lema *et al.*, 2014).

The German wind energy industry has its beginnings in the 1980s, when pioneers such as Enercon's Aloys Wobben developed Germany's first modern wind energy turbines and the first wind firms emerged. This was followed by a period of rapid upscaling of wind turbines from small kW turbines to MW turbines in the 1990s. These large turbines were called GROWIAN (GROsse WIndenergieANlagen – large wind energy turbines).

Nevertheless, the rapid upscaling of turbine capacities failed as the technologies tended to be sub-standard and unreliable and turbines broke down easily. As a result, the wind energy industry took a step back and developed smaller turbines, then improved and perfected them step-by-step, until the technology and the industry was sufficiently mature to gradually upscale the turbine capacities once again. At the end of the 1990s, the German wind energy industry grew rapidly thanks to new legislation – the so-called Renewable Energy Law (Erneuerbare-Energien-Gesetz, EEG) – introduced in 2000 by the Green–Red coalition government (the Social Democrats and the Environmental Party, in power from 1998 to 2005). Another important driver are feed-in tariffs (Lema *et al.*, 2014).

The wind energy industry in Germany is historically located onshore, with Enercon as the most prominent of the key German wind energy developers in terms of innovation and capacity, whereas the offshore industry has only recently begun to grow. Onshore wind technology in Germany is very advanced and has nearly reached the technical limit for energy efficiencies, according to wind energy experts. Turbine sizes fall into the multi-megawatt category, with Enercon's E-126 until recently the world's largest wind turbine with a capacity of 7.58 MW. Other experts argue that there is still potential for even larger turbines in the range of 10 MW. Offshore wind technology still needs more development with regard to grid access and grid extension issues, the installation of turbines, transport, (floating) fundaments, business models and materials (Lema *et al.*, 2014).

Wind energy capacity is unequally distributed in Germany. Most of the installed capacity is in northern Germany, close to the coast, where the wind is the strongest. The following federal states had the highest share of installed capacity at the end of 2011:

- Lower Saxony – 7,039 MW;
- Berlin/Brandenburg – 4,600 MW;
- Saxony Anhalt – 3,642 MW;
- Schleswig-Holstein – 3,271 MW;
- North Rhine-Westphalia – 3,070 MW (BWE, 2012).

Lower Saxony, the home of Enercon, has by far the highest installed wind energy capacity. Installed wind energy capacity in the southern federal states is small, but has been increasing in recent years, particularly in Baden-Wurttemberg and Bavaria, often in areas with low mountain ranges (*Mittelgebirge*). This is due to the fact that the windy areas in northern Germany have become increasingly saturated with wind energy farms, whereas there is still significant space for wind turbine developments in southern Germany. However, wind speeds are much lower in the south, so different turbines have to be used. Several firms, including Enercon and Vensys, have developed specific turbines for low wind speeds as a response to this challenge.

Another challenge in Germany is the imbalance between the southern federal states, which have low wind resources but high energy demand, and the northern federal states such as Lower Saxony and Schleswig-Holstein, which have strong winds and an oversupply of wind energy. The northern federal states request the improvement and expansion of grid infrastructure from north to south to facilitate long-distance electricity transport. The southern federal states, however, prefer to build more wind energy turbines in the south, despite low wind speeds, rather than import wind energy from the north. At the moment the political and geographical determinants have led to an expansion of energy capacity that is growing at a higher rate than the grid can handle. The greatest challenges are therefore grid extension for long-distance electricity transport from northern onshore wind farms to the south and grid integration and extension for offshore wind energy farms. Grid integration and grid expansion are key issues for the German wind energy industry as a large share of the wind energy has to be transported from the north to the south, which requires thousands of kilometres of grid extensions, which are often outdated and need improvement. Technically this should not be a major problem; hence grid-related challenges are mainly of a financial and political nature (BWE, 2012; Lema *et al.*, 2014).

Despite Germany's leading role in global wind energy, its wind energy industry remains understudied from an academic perspective. Existing studies cover important ground, but are mostly limited in terms of their geographic scope (for example, by focusing on one specific region in Germany) or because they adopt a narrow perspective (for example, by focusing on the public perception of wind energy or the cost of wind energy). For example, Arthur Jobert and colleagues (2007) and Fabian Musall and Onno Kuik (2011) discuss the local acceptance of wind energy in Germany; Michelle Portman and colleagues (2009) examine offshore wind energy by comparing Germany with the US; Martin Drechsler and colleagues (2012) focus on the feed-in tariffs in Lower Saxony in northern Germany; Russell McKenna and colleagues (2012) discuss the determination of cost–potential curves for wind energy in the federal state of Baden-Wurttemberg; and Marco Nicolosi (2010) discusses the wind power integration and power system flexibility with regard to extreme weather events in Germany.

Wind energy policy and its social implications

A central part of the low-carbon development discourse in particular and energy transition discourse in general has been a reliance upon technological change. There is therefore a growing discussion on how fossil fuels can be replaced by climate-friendly renewable energy, such as wind energy technology. Developing and deploying climate-friendly energy innovation systems is therefore portrayed as crucial to mitigating climate change.

These energy systems for climate change mitigation and low-carbon development can be referred to as sustainability-oriented innovation systems (Altenburg and Pegels, 2012). There has been a firm belief that government energy policies in Germany should be proactive in promoting low-carbon innovation. The German state is actively pursuing a strategy for a renewable energy transition instead of leaving this to the free market.

Energy transitions are shifts in a country's economic activities from an economy based on one energy source to an economy based (partially) on another energy source (Urban, 2014). Several energy transitions have occurred in history, mainly in developed countries:

- the transition from manpower and animal power to traditional biomass (such as fuelwood, crop residues and dung);
- the transition from traditional biomass to coal (*c.*1860);
- the transition from coal to oil (*c.*1880);
- the transition from oil to natural gas (*c.*1900);
- the transition from natural gas to electricity and heat (*c.*1900–1910);
- the large-scale commercial introduction of nuclear (*c.*1965);
- the large-scale commercial introduction of renewable energy and large hydropower (*c.*1995) (Bashmakov, 2007).

Energy transitions are characterised by changing patterns of energy use (for example, from solid to liquid to electricity), changing energy quantities (from scarcity to abundance or the other way around) and changing energy qualities (for example, from fuelwood to electricity) (Bashmakov, 2007). Bashmakov's three laws of energy transitions are as follows:

1 Energy transitions are often driven by changing energy costs in relation to income (for example, the predominant energy form becomes too expensive).
2 Energy transitions are often driven by improving energy quality (for example, the higher energy efficiency of electricity in comparison to fuelwood).
3 Energy transitions are often driven by growing energy productivity (for example, more industrial output can be obtained) (Bashmakov, 2007; Urban, 2014).

Low-carbon energy transitions should be understood as going from an economy based on fossil fuels to one based around low-carbon energy. The transition could be partial or more complete in both scope of sector and geography (Urban, 2014). Nevertheless, transitions and transformations are difficult as they need to be understood as part of a larger societal system. German wind energy policy should be understood as part of a larger national energy transition towards renewable energy and an attempt to front-load investment into renewables.

Berkhout and colleagues (2010) present the concept of *socio-technical regimes*, which describes "stable and ordered configurations of technologies, actors and rules that represent the basis for social and economic practices" and includes "a complex web of technologies, producer companies, consumers and markets, regulations, infrastructures and cultural values" (Berkhout *et al.*, 2010: 263). This is very much linked to the different development pathways that countries can take, which are constituted by a set of interlocking and interacting socio-technical regimes (Berkhout *et al.*, 2010). Energy systems could be described as "socio-technical configurations where technologies, institutional arrangements (for example, regulation, norms), social practices and actor constellations (such as user–producer relations and interactions, intermediary organisations, public authorities, etc.) mutually depend on and co-evolve with each other" (Rohracher and Späth, 2014: 1417).

There are some scholars who discuss how transitions and transformations are often countered by vested interests and perceived losses through new technologies and social change. This is often linked up to path dependencies and 'lock-ins': "Once a technological path has been embarked on, economies of scale and network externalities lead to reinforcing patterns which give incumbents a competitive advantage that makes switching to alternatives difficult" (Altenburg and Pegels, 2012: 12). Often this is considered in conjunction with fossil fuels as a 'carbon lock-in' (Unruh, 2000). We can here speak of the opposition of existing elites and the habits of populations and industries that prevents low-carbon transitions. Just as interesting as discussing what prevents a low-carbon energy transition is to discuss the social costs that such a transition will have on society. An ambitious energy policy will have to front-load some of the investment into low-carbon technologies and these costs will need to be carried by someone. The major social implications of ambitious wind energy policies are presented in Table 4.1.

The development of the wind energy industry in Germany has been reliant upon two key policies to promote intensive investments in wind energy: feed-in tariffs and a change in the building laws. Feed-in tariffs will be discussed below. The modification of the Federal Building Code in 1997 has made the rapid and uncomplicated development of wind energy possible. Since the building law was changed, wind turbines have been given a privileged status (Jobert *et al.*, 2007). Local authorities can "be forced to accept wind turbines on their territory" but they also have the power to assign "zones for wind energy farms, concentrating them on one appropriate site" (Jobert *et al.*, 2007: 2753).

The case of German feed-in tariffs

The boom in wind energy in Germany has been partly driven by national political decisions. This has been based on the political aim to phase out

Table 4.1 Environmental and social implications of wind energy

Environmental implications	*Social implications*	*Social sustainability*	*Social policy*
Wind energy is a low-carbon energy source, a climate-friendly alternative to fossil fuels and nuclear energy. It contributes to climate change mitigation and has limited direct environmental impact. There is some discussion around wind energy turbines and bird collisions. Environmental impacts are often discussed from an aesthetic point of view as blotting the landscape or obscuring views. There is also the need to upscale the grid system to be able to transport the energy from the source to areas where it is needed.	The direct social implication of wind energy is opposition to the impact on the human landscape (visually and/or through noise). A more urgent question is the increased cost of an energy transition that will be passed on to consumers and taxpayers and limit resources for others. A direct problem could be a rise in energy prices that will put pressure on the poorest.	Wind energy policies tend to focus on increasing energy output rather than improving the substantive and procedural aspect of wind energy policy; this would make the process more democratic and more focused on local communities and their needs. It is also important that the prices and costs are shared equally between corporations and the public.	The social implications of wind power, such as rising electricity costs, will have to be countered by increasing welfare support for the poorest. Those rising costs might lead to energy costs becoming a larger part of households spending. It is therefore important that the costs and risks are not covered solely by a rise in prices and that they are shared by taxes and corporations. It is also important that the benefits of wind power are spread to local communities.

nuclear energy by 2021 (BMU, 2011) and to replace both nuclear energy and fossil fuels with renewable energy. This is termed the 'energy transition' in Germany. The energy transition has been driven by the Renewable Energy Law (Erneuerbare-Energien-Gesetz, EEG) and the legislation for feed-in tariffs of 1991, 2000, 2004, 2009, 2010, 2012 and 2014. The EEG is seen as the cornerstone and key instrument for driving forward national innovation in wind energy in Germany. Table 4.2 gives details of the feed-in tariff subsidy.

About 98.5 per cent of this generating capacity in Germany comes from onshore wind energy, with only about 1.5 per cent of Germany's wind capacity currently coming from offshore wind energy (IEA, 2014). Table 4.2 shows that the EEG and feed-in tariffs have been instrumental in promoting the German wind energy sector. A lack of land means turbine sizes and capacities have to be upscaled and offshore developments have been favoured in recent years. This business innovation is driven by feed-in tariffs as they are paid per kWh of electricity generated, and are thus output-driven. A business model that increases the reliability of wind turbines to maximise financial gain from feed-in tariffs responds to the financial environment created by government policies (Nordensvärd and Urban, 2015).

An example of the impact of feed-in tariffs has been the rise of the German corporation Enercon. Enercon is a German wind turbine manufacturing firm founded in 1984. Enercon began by manufacturing small gearbox turbines, but shifted to gearless turbines in 1992. Today, Enercon has installed over 18,000 turbines worldwide. Enercon is Germany's most important and most established wind energy firm. It has a market share of over 60 per cent and has been operating for over 25 years (Urban *et al.*, 2015). Enercon and, to a lesser extent, other German wind energy firms have followed a distinctive wind energy innovation path. This innovation path focuses on high-quality, high-cost, large wind turbines predominantly using a direct-drive mechanism in place of gears and mainly operating onshore. Enercon is well known as the innovator that developed direct drive, a gearless technology (Nordensvärd and Urban, 2015). Enercon's

Table 4.2 German feed-in tariff for wind energy, 2014 version

Onshore wind	*Offshore wind*
First 5 years: 8.93 ct/kWh + 0.48 ct/kWh bonus = 9.41 ct/kWh After 5 years: 4.87 ct/kWh	Model 1: First 12 years: 15 ct/kWh After 12 years: 3.5 ct/kWh Model 2: First 8 years: 19 ct/kWh After 8 years: 0 ct/kWh

Source: adapted from Lema *et al.* (2014).

success can be linked to the feed-in tariffs, which indirectly encourage the upscaling of turbine sizes, outputs and projects. In 2007, Enercon launched its E-126 7.58 MW turbine, which was the world's largest wind turbine until 2014, when Vestas launched an 8 MW offshore turbine (Nordensvärd and Urban, 2015).

The German wind energy path has been shaped by decades of upscaling of turbines onshore while a recent boom in offshore operations is leading to a new growth phase in Germany's wind energy industry. The higher feed-in tariffs in the first years of a wind farm have attracted much investment in offshore wind operations. Nevertheless, these operations in Germany are costly and risky compared to the onshore sector. UNESCO's designation of the German Wadden Sea as a World Heritage Site has created natural restrictions on offshore developments along Germany's relatively short North Sea coastline. These restrictions mean that offshore wind farms have to be far off the coastline, which entails high investments and risks, and technological and logistical challenges, as well as challenges in terms of connecting the offshore farms with the national grid system (Nordensvärd and Urban, 2015).

Furthermore, in contrast to other countries that are leading the way with regard to offshore wind energy, such as Denmark, German offshore waters are deep, rough and difficult to manage. This has meant dependence upon large multinational utility companies such as E.ON and Vattenfall to be able to provide the large investment sums and to co-ordinate the increasingly complex task of building and maintaining the wind farms (Nordensvärd and Urban, 2015). An example of this is the Alpha Ventus project. Alpha Ventus was the first German offshore wind farm. It was essentially a test field that was built to gain experience for the German offshore wind industry. It was built in the North Sea, 45 km offshore, in proximity to the East Frisian island Borkum (Alpha Ventus, 2012a, 2012b). Second-generation offshore projects, such as Vattenfall's Dan Tysk and EWE's Riffgat, have built upon lessons learnt from the Alpha Ventus experience. Dan Tysk operates in even deeper waters (70 km offshore) and even rougher weather conditions. In total, 11 offshore wind farms have been built in Germany at the time of writing.

The German energy transition and feed-in tariffs have created their own path dependency and technological lock-in. This could be considered as a feed-in tariff lock-in, which has led to path dependency towards upscaling of wind energy turbines and projects. It has led to larger wind energy turbines and projects onshore as well as more financially risky offshore projects, since these tend to produce larger outcomes per megawatt, and hence receive higher subsidies. Large corporations, their financial resources and their experience have become more important in the growing offshore wind industry. In a sense, Germany has become a financial and technological laboratory for offshore wind energy; this is due to the rapid development of offshore wind energy with larger turbines, which have often not been tested or installed elsewhere, and installations further out at sea. The costs of these risks and large investments are not only covered by feed-in tariffs, but also by higher energy prices for consumers (Nordensvärd and Urban, 2015). These higher prices are however not due to rising energy prices caused by renewables; on the contrary, energy prices have declined as more renewables have been introduced in the energy mix in Germany. Instead, the higher energy prices for consumers have been created by an unfortunate reform of the Renewable Energy Law (EEG) in 2010 and by large energy firms that pass price increases on to their customers but keep profits for themselves (IWR, 2014a). The social dimension of feed-in tariffs has had a direct impact on sharing the burden of the transition towards a low-carbon economy.

Substantive dimension

German feed-in tariffs have become the driver for the ever-expanding creation of new wind energy capacity. The sector is thriving due to state subsidies, but there is a shrinking political will to maintain the costly feed-in tariff system in its current form. A recent appraisal by the Expert Committee for Research und Innovation (EFI) concluded that the EEG and the feed-in tariff policy have led to a sharp increase in renewable energy, from 7 per cent of the total electricity supply in 2000 to 23 per cent in 2012 (EFI, 2014). At the same time, the costs paid by taxpayers and the government to the operators/energy providers have increased from €1.6 billion in 2001 to €22.9 billion in 2013 (see Figure 4.1). The EFI argues that the

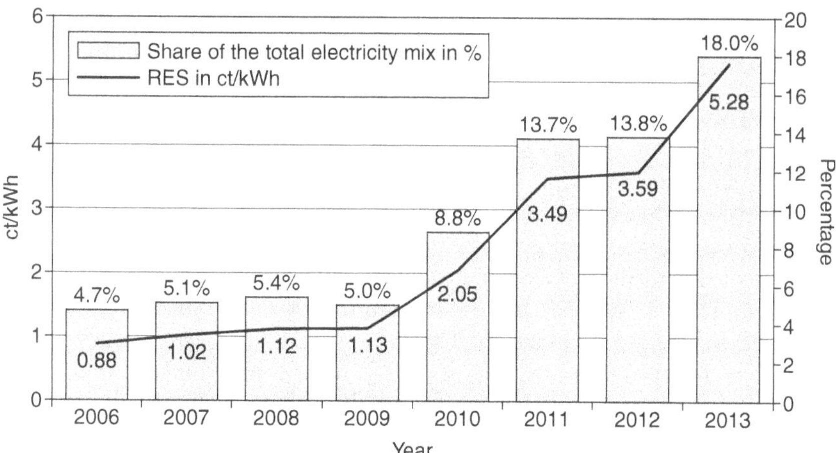

Figure 4.1 Renewable energy surcharge in ct/kWh based on feed-in-tariff and renewable energy as a percentage share of the total electricity mix in Germany, 2006–13.

Source: EFI (2014).

Table 4.3 Feed-in tariffs for wind energy in Germany

Substantive dimension	*Procedural dimension*	*Social policy dimension*
It is argued that consumers and taxpayers have paid through feed-in tariffs for the expansion of renewable energy in general and wind energy in particular. Others argue that the increase in prices is mainly due to the way that the EEG was reformed and due to the greed of large energy corporations (IWR, 2014a). The focus on upscaling has led to the involvement of large multinational corporations that build costly offshore wind farms. The main issue is whether consumers should sponsor large profit-driven corporations that have led to higher energy prices.	The German approach has followed an industrial model that promotes the upscaling of turbine sizes and capacities and large offshore wind farms. The modification of the Federal Building Code has removed the possibility of opposing the construction of wind energy sites. Most interesting is the opposition against a large electricity link (*Stromtrasse*) between the north and the south. This calls for wind energy demand to be met locally instead of the long-distance transport of electricity between the north and the south. The system is currently developed for the benefit of large utilities.	Feed-in tariffs have shifted the costs partly away from corporations and the state towards consumers. Energy-intensive industries continue to pay a lower rate for renewable-based electricity than consumers, which could be seen as corporate welfare. Moreover, when prices have dropped on the energy markets, this has not been passed on to consumers. The financial winners are electricity buyers and the industry (IWR, 2014a). Cynics could therefore argue that the German wind energy policy is promoting corporate welfare at the expense of consumers.

sharp increase in electricity production from renewable sources has led to a sharp rise in costs for taxpayers and consumers. The EFI reports that by 2014 one-fifth of Germany's average energy costs was due to feed-in tariffs. This has led to a critical public discussion on the legitimacy of the EEG and feed-in tariffs (EFI, 2014: 2). However, the real question is whether subsidies for wind energy in Germany need to be formulated more sustainably so that innovation is encouraged and firms can flourish without excessive costs for consumers. Currently, the focus on upscaling wind turbine sizes and projects while neglecting other forms of innovation such as grid systems or smaller turbines has led to a socio-technical system characterised by inertia and lock-in. This means that the German nation state per se has difficulties in creating an environment that stimulates innovation in renewable energy but that does not result in rising costs for consumers.

One of the main issues surrounding feed-in tariffs is who is going to shoulder the major costs of the investment in low-carbon technology. Feed-in tariffs mean that renewables are subsidised by a surcharge on the price of electricity, which means that consumers have been asked to pay for renewable energies. This has led to a rise in costs over recent decades. The total subsidy is currently about €20 billion a year, which amounts to €218 a year per household on top of the normal electricity bill (Brunekreeft, 2014).

The cap in 2014 on feed-in tariffs has led to an average of 6.17 ct/kWh for 2015, which is a small decline compared to 2014 (Brunekreeft, 2014). The feed-in tariffs for renewable energy (*EEG-Umlage*) are not funded by the state, nor are they subsidised through public funding or taxation. Similar to the minimum wage, the government sets the absolute minimum price for electricity from renewable sources. The operators of renewable energy sites such as wind farms receive a specific sum in return for the feed-in electricity. This electricity is then sold on the stock market to generate profits. Feed-in tariffs cover the difference between revenues and costs (IWR, 2014a). The major challenge is that even since feed-in tariffs for renewable energy have been capped, this has not resulted in falling prices; prices are actually rising. This can be explained by the fact that the government is exempting parts of the industry from feed-in tariffs, which meant a rise of costs from €1 billion to €5 billion by 2014. Corporations that use large amounts of energy are only paying 0.05 ct/kWh for the feed-in tariff (virtually an exemption) compared to 6.17 ct/kWh for private households. The energy consumption threshold at which exemption is granted has been lowered from 10 GWh to 1 GWh, which means that payments into the EEG account have decreased and the costs of the transition have to be shouldered by consumers (IWR, 2014a). Ironically, the way that the EEG and feed-in tariffs were reformed in 2010 has led to consumers being the economic losers while large firms, utilities and large energy providers are the economic winners of the energy transition towards renewable energy in Germany. An increased share of wind energy and other renewables has indeed lowered energy prices, albeit only for large corporate customers and energy providers, not for individual consumers.

The rise of renewable energies on the stock market led to lower prices, which was down to a change in the regulations, whereby suppliers had to serve municipal utilities directly with renewable energy; this changed in 2010, when renewable energy had to be sold on the stock market. Municipal utilities are now supplied with conventional electricity and since more renewable energy is sold on the stock market, this also means a drop in prices (IWR, 2014a). The reduction in energy prices has led to large consumers and industries profiting, but the reduced cost of energy has not been passed on to consumers as most electricity providers do not buy electricity on the stock market but do so using more secure full-supply contracts according to fixed prices. The cost of the changes to the EEG are then added on top of the consumer price (IWR, 2014a).

Procedural dimension

German wind energy policy has not only led to a lock-in that supports only large turbine design and upscaling energy generation output, but also a system that supports economic actors (the producers and suppliers of renewable energies on the one hand and energy-intensive industries on the other) on the back of individual household consumers. The system shows that the average household shoulders a disproportionate share of the costs but does not have a say in how the resources are spent. Most of the energy that consumers buy (albeit marketed otherwise) is from conventional energy sources (IWR, 2014a).

In addition, feed-in tariffs create their own path dependency, which has excluded other innovation paths for low-carbon development in general and wind energy in particular. This finding is supported by other studies, which argue that small wind energy turbines in Germany have received little attention and fewer subsidies through feed-in tariffs and the EEG (BWE, 2010).

Alternatives to use wind energy more efficiently through smart grids, citizens' networks or electro-mobility have not been sufficiently pursued. "[T]he French and Chinese governments subsidise the purchase of electric vehicles to kick-start production, assuming that, once certain minimum scales of production are reached, the price gap between these vehicles and conventional cars will narrow considerably" (Altenburg and Pegels, 2012: 12). This has not happened in Germany, as the participation of citizens in the decision-making process has been limited and there has been a greater focus on large corporations.

The main issue is that northern Germany, including the offshore wind farms, produces more wind energy than the ageing grid system can handle. Hence, the electricity from the north is not distributed efficiently to the south, where the demand for electricity is high. The plans to build an electricity link (*Stromtrasse*) between northern Germany (Niedersachsen, Schleswig-Holstein) and southern Germany (Bavaria, Baden-Wurttemberg, Hessen) have been met with enormous public and political resistance. The federal state of Bavaria has requested a moratorium on the project. Furthermore, many civil society organisations are planning protests against the project (Nordensvärd and Urban, 2015).

The overall judgement is that the system supports large suppliers and providers in a way that does not focus on the participation and involvement in decision-making of households, communities and citizens. The accountability towards consumers is not transparent as most consumers do not know what energy they pay for or why energy costs are rising. Even if the German system has been successful in ramping up the production of wind energy, this does not mean that it has followed a socially just path. Moreover, this shows a path dependency in German energy policy to supply the industry with incentives to invest and gain profit from renewable energy.

Social implications, challenges and solutions

The major social implications of German wind energy policy are that consumers have been bearing the brunt of the costs while corporations have been gaining the profits from the wind energy transition. Instead of contemplating raising money for a front-loaded investment in renewable energy, the German government focused on a model that raised the costs of consumers' energy bills without really improving their participation in renewable energy such as wind power. The German wind energy sector has been highly dependent upon political decisions taken since the early 1990s to support a fast and ambitious expansion of wind energy. As part of a government reshuffle in late 2013 (resulting from the electoral victory of the so-called 'Grand Coalition'), the entire energy agenda was transferred to the Ministry of Economic Affairs and Energy. This includes overall energy policy, grid management and the promotion of renewable sources of energy as well as energy efficiency enhancement. Importantly, the management and reform of Germany's energy transition is thus directly linked to a comprehensive agenda of economic growth, competitiveness and innovation.

The tendency to reduce electricity prices for German industries with high energy intensity is a truly divisive strategy as this means that households are more or less subsidising costly and maybe also unsustainable industries through their energy bills. This could be seen as problematic from many different perspectives.

First of all, one could argue that this is a form of corporate welfare whereby households more or less support corporations that use large amounts of energy. Corporate welfare is often defined as any action taken by the government that provides benefits to a corporation or industry not offered to others (Barlett and Steele, 1998); in this case, the German state gives preferential treatment to classical industrial corporations that are energy intensive. This means that other industries with lower energy use or from other countries are disadvantaged. Most critics of corporate welfare have found that there has been little corresponding return to taxpayers (Antonelli, 1995; Barlett and Steele, 1998; Moore and Stansel, 1995) and programmes favoured one company over another (Lindsey, 1992; Schatz, 1997). In this case, one could argue that energy consumers have not really benefited from the exemptions given to industries by the German government.

Second. this should also be understood as giving German corporations an advantage without using subsidies or tax credits. From a historical perspective, the EU has tended to be seen as a barrier that sought to restrict the development of the EEG in the first place because the law encouraged the preferential treatment of one member state over another and encouraged 'unfair competition' within the EU. The German policies of the EEG are much more ambitious than the EU policies; for instance, the EU 2020 target is 20 per cent renewable energy of the total installed energy capacity, whereas the German target is 35 per cent and by 2011 almost 20 per cent of renewable energy had been installed (Lütkenhorst and Pegels, 2014). Feed-in tariffs are in practice the preferential treatment of energy-intensive German industries. Even if the German renewable energy policy was to create a barrier to the single market, this would be overridden by the prioritisation of climate protection and the interest in supporting renewable energy. In this case, the EU supports Germany's transition to renewable energy but takes issue with its preferential treatment of its own industries. The problem that the EU has is that corporations that use large amounts of energy are only paying 0.05 ct/kWh for energy compared to 6.17 ct/kWh for private households. The EU Commission argues that exempting these industries from feed-in tariffs should not be seen as a climate protection policy but actually as a subsidy and the preferential treatment of German industries (IWR, 2014a).

Third, it is ironic in a sense that feed-in tariffs were created by the coalition government of the Social Democrats and the Green Party, but then amended by the Conservative–Liberal government in 2010 into their current socially unsustainable form. The present government's actual wind energy policy is quite far removed from the original idealism of the Green Party. The Green Party actually started the trend towards wind energy back in the early 1980s, in opposition towards both nuclear power and fossil fuel energy. The green movement also includes various environmental non-governmental organisations (NGOs), most notably Greenpeace, that have been opposing nuclear energy as well as fossil fuel energy in Germany for decades in favour of renewable energy sources such as wind. Unlike in many other countries, the green movement is powerful in Germany and the Green Party grew in significance in the late 1990s and the 2000s and even came to power in a coalition with the Social Democrats. Today, the Green Party forms part of the government in several federal states, such as in Baden-Württemberg and North Rhine-Westphalia, and drives a strong political agenda that favours renewable energy, particularly wind energy. The main issue is here that the government, including the Green Party, has accepted a system that favours large-scale, corporate and industrial use of energy. It is far from supporting local communities or small-scale and citizen-managed wind farms. The system has favoured centralised solutions whereby most wind energy is produced in the north, but on the other side there have been no comprehensive plans as to how the energy should be transported from the north to the south, where the energy is mostly needed.

There is also the practical problem of connecting the wind farms to the grid, which is complicated for offshore wind farms. The grid provider will only establish a connection when it is guaranteed that the wind farm will be successfully completed. For offshore wind farm providers, this is a catch-22 as they need confirmation from grid providers before being able to secure funds from investors. A new law on exemption from liability (*Haftungsausschluss*) requires that the risk is shared between wind farm operators, grid providers, the public/taxpayers and consumers. Hence some of the burden of offshore wind farms is at the expense of taxpayers/consumers (Nordensvärd and Urban, 2015).

Some argue that it is the responsibility of the government to push forward grid integration and grid expansion. They argue that government authorities need to regulate the actors more tightly and force grid operators such as Tennet to invest in expanding the grid and making it compatible with offshore wind energy. Some interviewees point out that large offshore projects such as Alpha Ventus have been delayed for two years due to grid connection issues. There is also an agreement that federal states need to be involved so that electricity transport from the north to the south will be smooth and efficient. A challenge at the moment is that southern federal states want to become more independent and increase their own wind energy capacity. In southern Germany, turbines are being built in low-wind-speed areas, which could decrease the demand for wind energy from the north. The state could therefore be understood as the main determinant of the German wind energy innovation path, which means that the growth of the industry is very much dependent upon the political dimension of wind energy and how the energy transition is financed. This means a path dependency towards large offshore wind farms that have long-term problems in getting connected to the grid and an outdated transmission and distribution network that fails in delivering the electricity to where it is most needed.

The question is whether consumers should also pay for the building of large controversial energy links. Some of the major challenges facing feed-in tariffs can be found in Table 4.4.

The previous coalition government (Conservatives–Liberals: CDU/CSU and FDP) and the current coalition government (Conservatives–Social Democrats: CDU and SPD) have discussed ways to reduce feed-in tariffs. The EFI report confirms that the EEG and feed-in tariffs have led to an expansion of renewable energy in Germany, but doubts whether innovation in renewables has been sufficiently stimulated through the schemes. The report's radical, contentious and contested recommendation is to abolish feed-in tariffs altogether (EFI, 2014: 2). However, this is strongly contested by wind energy firms, which argue that they and the entire wind energy sector would go bankrupt if feed-in tariffs were to be abolished or significantly reduced. One could argue that the current system has become a "stable and ordered configuration of technologies, actors and rules that represents the basis for social and economic practices" and includes "a complex web of technologies, producer companies, consumers and markets, regulations, infrastructures and cultural values" (Berkhout *et al.*, 2010: 263). It has become the dominant approach in Germany that consumers should pay for the energy transition while large energy users such as energy-intensive industries and energy

Table 4.4 Social implications, challenges and solutions

Social implications	*Challenges*	*Solutions*
The problem is that polluters and corporations have not paid the lion's share of the costs for the government's wind energy policy. The controversial part is that household electricity bills have been rising without compensation from the state or energy providers, while wholesale electricity prices have been decreasing for many years thanks to the increasing share of renewables. This means that the direct social implication is rising household electricity bills in particular and poorer households in general. At the same time, large industries and energy providers have benefited financially from decreasing wholesale prices due to the increasing share of wind energy and other renewables among the German energy mix.	The major challenge is to be able to muster the large front-loaded investment for wind energy without consumers and taxpayers having to pay for the major part of the energy transition. There is a danger that the German system might erode the public support for energy transition in general and wind energy in particular.	There is a lack of discussion on how the costs of the energy transition in Germany could be funded through, for example, higher taxes for polluting industries or taxing high industrial use of electricity. Germany has taken the corporate, large-scale route, thereby neglecting smaller wind energy turbines that receive fewer subsidies through the feed-in-tariff and the EEG (BWE, 2010).

operators and suppliers reap profits from the transition, on the back of individual consumers.

In conclusion, Germany is indeed a country that is one of the world leaders with regard to wind energy innovation, wind energy installed capacity and wind energy markets and a front-runner in the renewable energy transition. Germany sets an example for climate change mitigation. However, what is good for the climate may not always be good for people and what is environmentally sustainable may not always be socially sustainable. Germany has failed to make the energy transition socially just and equitable. Instead, it has reformed a policy that was hailed worldwide and transformed it into an instrument that mainly creates profits for wealthy industries and energy operators while increasing financial disadvantages for households and individual consumers. As is so often the case, the poor are hit the hardest as they cannot keep up with the rising energy bills. This is not only socially questionable, but also gives renewable energy a bad name and risks the credibility of the energy transition. This serves as another reminder of how low-carbon development policy needs to be carefully planned and implemented from an environmental, social and economic perspective.

Acknowledgement

This chapter is taken and adapted from Nordensvärd and Urban (2015), with kind permission from Elsevier B.V.

References

Alpha Ventus (2012a), *Alpha Ventus: Technical Report*, Oldenburg: DOTI.

Alpha Ventus (2012b), *Alpha Ventus: Executive Summary*, Oldenburg: DOTI.

Altenburg, T. and Pegels, A. (2012), "Sustainability-oriented innovation systems: managing the green transformation", *Innovation and Development*, 2(1), 5–22.

Antonelli, A. (1995), "Congress, not Clinton, supports the end of corporate welfare", Backgrounder Update No. 253, Washington, DC: Heritage Foundation.

Barlett, D. L. and Steele, J. B. (1998), "Special report: exposing the folly of corporate welfare", *Time*, 152(19), 9 November.

Bashmakov, I. (2007), "Three laws of energy transitions", *Energy Policy*, 35(7), 3583–94.

Berkhout, F., Verbong, G., Wieczorek, A. J., Raven, R., Lebel, L. and Bai, X. (2010), "Sustainability experiments in Asia: innovations shaping alternative development pathways?", *Environmental Science & Policy*, 13(4), 261–71.

BMU (German Federal Ministry for the Environment) (2011), "Background information for the expansion of renewable energy in Germany by 2020" (in German), available at: www.bmu.de/files/pdfs/allgemein/application/pdf/hintergrund_ausbau_ee_bf.pdf (accessed 22 March 2016).

BMU (German Federal Ministry for the Environment) (2012), "Renewable Energy Law EEG – 2012 version" (in German), available at: www.bmu.de/files/pdfs/allgemein/application/pdf/eeg_2012_bf.pdf (accessed 22 March 2016).

Brunekreeft, G. (2014), "Germany's green power surge has come at a massive cost", *The Conversation*, 20 October, available at: https://theconversation.com/germanys-green-power-surge-has-come-at-a-massive-cost-33202 (accessed 22 March 2016).

BWE (German Wind Energy Association) (2010), *Cost-Effectiveness and Subsidies for Small Wind Energy Turbines* (in German), Berlin: BWE, available at: www.wind-energie.de/sites/default/files/download/publication/wirtschaftlichkeit-und-vergutung-von-kleinwindenergieanlagen/bwe_kwea_studie_liersch_final_2.pdf (accessed 22 March 2016).

BWE (German Wind Energy Association) (2012), *Wind Industry in Germany, 2012*, Berlin: BWE.

Drechsler, M., Meyerhoff, J. and Ohl, C. (2012), "The effect of feed-in tariffs on the production cost and the landscape externalities of wind power generation in West Saxony, Germany", *Energy Policy*, 48, 730–6.

EFI (Expert Commission on Research and Innovation) (2014), *Report on Research, Innovation and Technological Efficiency in Germany* (in German), Berlin: EFI.

GWEC (Global Wind Energy Council) (2014), *Global Wind Report: Annual Market Update 2013*, Brussels: GWEC, available at: www.gwec.net/wp-content/uploads/2014/04/GWEC-Global-Wind-Report_9-April-2014.pdf (accessed 22 March 2016).

IEA (International Energy Agency) (2014), "Energy statistics", available at: www.iea.org/stats/ (accessed 22 March 2016).

IWR (German Institute of the Renewable Energy Industry) (2014), "Renewable energies are subsidised – the state doesn't pay a cent" (in German), Münster: IWR, available at: www.iwr-institut.de/de/presse/presseinfos-energiewende/erneuerbare-energien-werden-subventioniert-staat-zahlt-keinen-cent#wie-die-EEG-Umlage-funktioniert (accessed 22 March 2016).

Jobert, A., Laborgne, P. and Mimler, S. (2007), "Local acceptance of wind energy: factors of success identified in French and German case studies", *Energy Policy*, 35(5), 2751–60.

Lema, R., Nordensvärd, J., Urban, F. and Lütkenhorst, W. (2014), "National innovation paths in wind power? Insights from Denmark and Germany", Working Paper. Bonn: GDI.

Lindsey, B. (1992), "Sematech: the wrong solution", *Journal of Commerce*, 16 January.

Lütkenhorst, W. and Pegels, A. (2014), *Stable Policies, Turbulent Markets: Germany's Green Industrial Policy – the Costs and Benefits of Promoting Solar PV and Wind Energy*, Ottawa: International Institute for Sustainable Development.

McKenna, R., Gantenbein, S. and Fichtner, W. (2012), "Determination of cost-potential curves for wind energy in the German federal state of Baden-Württemberg", *Energy Policy*, 57, 194–203.

Moore, S. and Stansel, D. (1995), "Ending corporate welfare as we know it", Cato Policy Analysis No. 225, Washington, DC: Cato Institute.

Musall, F. D. and Kuik, O. J. (2011), "Local acceptance of renewable energy: a case study from southeast Germany", *Energy Policy*, 39(6), 3252–60.

Nicolosi, M. (2010), "Wind power integration and power system flexibility: an empirical analysis of extreme events in Germany under the new negative price regime", *Energy Policy*, 38(2), 223–34.

Nordensvärd, J. and Urban, F. (2015), "The stuttering energy transition in Germany: wind energy policy and feed-in tariff lock-in", *Energy Policy*, 82, 156–65.

Portman, M. E., Duff, J. A., Köppel, J., Reisert J. and Higgins, M. E. (2009), "Offshore wind energy development in the exclusive economic zone: legal and policy supports and impediments in Germany and the US", *Energy Policy*, 37(9), 3596–607.

Rohracher, H. and Späth, P. (2014), "The interplay of urban energy policy and socio-technical transitions: the eco-cities of Graz and Freiburg in retrospect", *Urban Studies*, 51(7), 1415–31.

Schatz, T. A. (1997), "Prepared testimony of Thomas A. Schatz, president, Citizens Against Government Waste, before the Senate Committee on Governmental Affairs", Washington, DC: US Senate.

Unruh, G. C. (2000), "Understanding carbon lock-in", *Energy Policy*, 129(1), 817–30.

Urban, F. (2014), *Low Carbon Transitions for Developing Countries*, London: Routledge.

Urban, F., Zhou, Y., Nordensvärd, J. and Narain, A. (2015), "Firm-level technology transfer and technology cooperation for wind energy between Europe, China and India: from North–South to South–North cooperation?", *Energy for Sustainable Development*, 28(10), 29–40.

5 Social implications of carbon markets

The case of carbon offsets and Plantar in Brazil

Johan Nordensvärd with Tamra Gilbertson and Domiziana Marinelli

Carbon markets and their social implications

Carbon markets have become a dominant discourse in tackling climate change. The UN continues to focus on the importance of carbon trading and offsets as primary policy tools. There have been various implementations of carbon trading. The largest carbon trading scheme is the EU Emissions Trading System (EU ETS), a cap-and-trade mechanism that sets a cap on greenhouse gas emissions and issues permits to polluting industries according to these caps. Once emission allowances are exceeded, either emission permits must be bought from those who have emitted less or companies must purchase offsets. It is perceived that carbon emissions are not just a cost that needs to be passed on to producers and consumers, but also a cost that could be traded and become an integral part of a global market. Therefore carbon trading is perceived to be a market-friendly system as it is not based around tax or command-and-control regulations. There is an argument that a market for offsets incentivises entrepreneurs to seek out new ways of reducing their own emissions so that they can then trade offsets, but there is a risk that these reductions could be cancelled out by the additional emissions they supposedly 'compensate' for elsewhere.

In order to comprehend their impacts, carbon markets need to be understood not just as a particular policy tool but also as a product of a particular economic perspective. The neoclassical explanation of climate change is that emissions exist in the economy as externalities that are not reflected in the exchange of goods on the market. "To deal with such externalities, new methods of environmental valuation and market-based solutions to protect the environment have been introduced across the world" (Pearse and Böhm, 2014: 326). Carbon markets are seen as the opposite of command-and-control regulation, which establishes standards, taxes or legislative bans on harmful substances and practices. Some see the development of carbon markets as being linked with the general ideological turn towards neoliberalism and the pursuit of market and quasi-market solutions (Pearse and

Böhm, 2014). This reframes the role of the state as being to set emission caps, provide the infrastructure to trade, collect monitoring reports and solve conflicts and judicial issues. The market is then largely left to sort out the pricing and the global trading mechanisms (Scott, 2008). The Stern Review argues that carbon markets are part of a "pro-growth strategy" (Stern, 2006: iii), which should offer financial returns for investors and create a productive global space for low-carbon technologies. As carbon trading has become part of the mainstream discourse, the IEA (2008), the IMF (2008) and the World Bank (2008) continue to incentivise the market mechanisms to cope with climate change. The UNDP argues that aiming to avoid massive quotas and bureaucratic controls rather than focusing on adjusting the price system is advantageous (UNDP, 2007: vii). However, even if state or public organisations have to play extensive roles in a global carbon scheme, this pathway is still seen by some as an 'alternative' compared to carbon taxation or a command-and-control approach (Milner, 2007). The UNDP argues that "carbon markets are a necessary condition for the transition to a low-carbon economy" (UNDP, 2007: 11).

Larry Lohmann argues that the commodification of nature is far from a new phenomenon:

> First proposed in the 1960s, pollution trading was developed by US economists and derivatives traders in the 1970s and 1980s and underwent a series of failed policy experiments in that country before becoming the centrepiece of the US Acid Rain Programme in the 1990s at a time of deregulatory fervour.
>
> (Lohmann, 2010: 78)

The Clinton administration advocated carbon trading as the main mitigation mechanism of the Kyoto Protocol (Lohmann, 2010).

There is a need to have a global regime to be able to facilitate a global carbon market, and this faces several challenges, as discussed by Lohmann. First, governments have to determine "how much of the world's physical, chemical and biological ability to regulate its own climate should be enclosed, 'propertised', privatised and made scarce" (Lohmann, 2010: 80). The state distributes or sells these pollution rights to large polluters before the market decides how these will be distributed (Lohmann, 2005, 2006). For the carbon market to function, environmental goods and bads need to be turned into quantifiable items that can be exchanged. There is the perception that the market will somehow find the places where emissions reductions could happen with the least cost. It is also important to create compensatory projects so carbon emission in industrialised countries can lead to low-carbon development in low- and middle-income countries; these projects take the form of offset credits. Finally, it is necessary to create institutions and regulations to facilitate the commodification, by selling and offsetting carbon emissions (Lohmann, 2010). Lohmann argues that a commodity approach "abstracts from where, how, when and by whom the cuts are made, disembedding climate solutions from history and technology and re-embedding them in neoclassical economic theory, trade treaties, property law, risk management and so forth" (Lohmann, 2010: 81). In practice, pollution grants are primarily for industrialised countries, firms and other polluters, which often receive generous pollution rights to appease possible opposition.

An important aspect of the commodification process of carbon emissions comes down to some key questions: who owns the emissions, who owns the trees and land, and what happens when the carbon emissions prices nose-dive. The Bolivian president Evo Morales accused the developed world of "not only ... cheat[ing] their commitments to reduce emissions, but ... also [beginning] the commoditization of nature.... We cannot accept ... any mechanism of carbon markets or 'incentives' that may lead to the commodification of forests and rainforest" (Morales, 2010 cited in Lohmann, 2011: 657).

Defenders of carbon trading tend to focus more on the perceived positive that a more market-friendly system can be more cost effective. Stern argues that emissions trading systems "can deliver least cost emissions reductions by allowing reductions to occur wherever they are cheapest" (Stern, 2007: 326). Carbon trading is often perceived to be the preferred mechanism by corporations and a less costly alternative to a carbon tax (Stern, 2009). Jonathon Porritt has argued that a carbon trading quota will "clearly have a huge impact upon the business community, directing companies into ever more ingenious ways of getting the carbon out of whatever it is that they are selling" (Porritt, 2005: 237–8).

A core element in the Stern Review's argument is that technologies and financial flows from carbon trading can for example take care of the 95 per cent reduction in carbon intensity that it sees as necessary for China's industry (Lohmann, 2011: 657). Even scholars who are critical of limitless economic growth, such as Herman Daly, have jumped on the carbon market bandwagon. Daly has argued that "the tradable permits idea can be applied to limit the greenhouse effect ... [and] is truly a paradigm for many sensible policies" (Daly, 1996: 56). Furthermore, he continues, the tradable permits plan is "the paradigm policy for solving the allocation, distribution, and scale problems" (Daly, 1996: 223–4). One of Daly's most recent publications (Daly, 2015) suggests that his views might have changed.

The EU ETS is the largest emissions trading system in operation in the world and has been subject to fraudulent activity and market volatility, as well as rewarding some of the biggest polluting industries with windfall profits (Reyes, 2011). Despite its failures, many other existing and prospective markets have been modelled on the EU ETS. Since 2013, new carbon markets have been introduced in California, Kazakhstan, Mexico, Quebec, Korea and China. Although there is interest from national and international bodies in linking emissions trading systems, many of these markets are subject to political and economic uncertainties, at best. There has been a noticeable critique of carbon markets as being the wrong answer to the wrong question. Rebecca Pearse and Steffen Böhm (2014) and Clive Spash (2009) have explored some of the most obvious deficiencies of carbon markets. This chapter will discuss a particular aspect of carbon trading which has become a dominant feature: carbon offsetting.

Attempts to implement carbon markets

Cap-and-trade systems often rely upon offsets to function, but there are exceptions such as the US sulphur dioxide program, which had no offsets. Offsets provide a way to pollute beyond the agreed cap with respect to individual firms, as do purchased allowances (no offsets needed); it is only with respect to a non-global cap that offsets are needed to go beyond the agreed cap by purchasing credits from supposed technology transfer or preventing deforestation. The whole concept of offsetting emissions was built into the Kyoto Protocol:

> The concept of 'offsets' was created under the Kyoto Protocol to refer to emissions reductions not covered by the cap in an ETS. A standard permit system requires a seller to have controlled their source emissions to be able to sell a permit. Offsets are based upon projects which are disassociated from the polluting source and either reduce GHG emissions elsewhere or increase the capacity of a sink (such as forests or soils) to absorb GHG pollution (such as carbon), beyond 'business as usual'. Offsets are also now widely traded outside the Kyoto-compliance market, including by individuals and firms voluntarily aiming to offset their GHG emission.
>
> (Spash, 2009: 48)

Offset credits can be sold to polluting industries that have obligations to reduce their emissions in countries regulated by cap and trade, and contribute to the 'flexibility' in terms of where reductions will be undertaken. Offsetting means that polluting actors do not have to reduce their emissions but can actually continue to increase emissions as long they offset them elsewhere.

The largest offset mechanism is the Clean Development Mechanism (CDM), which enables developed countries to buy credits from projects in developing countries that reduce or capture emissions. In theory, developing countries gain access to climate-friendly technology, such as renewable energy, while developed countries gain Certified Emission Reduction credits (CERs) to offset their emissions. By early 2012, 3,742 projects had been registered and another 7,500 were in the pipeline (UNEP, n.d. a). By October 2015, the number of registered CDM projects had doubled to 7,771, 91 per cent of the 8,541 projects in existence at that time (UNEP, n.d. a).

The large majority of CDM projects are located in emerging economies, most of which are in China and India. The CDM is currently being reformed to enable least-developed countries (LDCs) to benefit from the mechanism, as poor countries – particularly in Africa – have attracted little funding through it. Another key criticism of the CDM is that it is an offsetting mechanism. This means that rather than reducing emissions, countries and firms invest in projects in poor countries. This could be described as a modern form of selling indulgences (compare Urban and Nordensvärd, 2013).

Another example of an offset mechanism is joint implementation (JI), whereby developed countries can invest in CER projects in other developed countries as an alternative to reducing emissions domestically. JI is currently in place for developed countries only. For example, Eastern European countries, where carbon emissions have declined since the fall of the Soviet Union, can reduce their emissions jointly with other countries that have growing emissions, for example Japan. Most JI projects are implemented in the energy efficiency and methane reduction sectors. By early 2012, a total of 305 projects with 143,438 emission reduction units (ERUs) had been recorded (UNEP, 2012b). By October 2015, there were 863 projects (UNEP, n.d. b). Current research has shown that "faults in JI have released 600 million tons of carbon dioxide more than if the scheme wasn't in place" (Kollmuss *et al.*, 2015).

A third example of an offsetting strategy is the Reducing Emissions from Deforestation and Forest Degradation (REDD+) mechanism, which was introduced as a new mechanism in the Copenhagen Accord (UNFCCC, 2009) and the Cancun Agreements aimed at operationalising REDD+ (UNFCCC, 2010). Under REDD+, developing countries can be paid for forest management, conservation and enhancement of forest carbon stocks, while developed countries can gain CERs to offset their emission obligations. Major initiatives include the Forest Investment Programme, the Forest Carbon Partnership Facility and the UN-REDD Programme. The UN-REDD Programme estimates that US$30–100 billion per year can be brought into developing countries as payments for REDD+ (UN-REDD, 2010). Recipients of REDD+ payments are developing countries with tropical forests, including countries that have high deforestation rates such as Brazil, the Democratic Republic of Congo (DRC) and Indonesia. REDD+ has been heavily criticised by Indigenous and forest-dependent communities as a mechanism to create more land enclosures and to control patterns of forest use and restrict land rights. Furthermore, REDD+ has in some cases led to land grabs in Africa, Indonesia, Brazil and Papua New Guinea.

The clear social dimensions embedded within offsetting strategies are often overlooked; however, emissions are historically the responsibility of wealthy industrial nations. Carbon offsetting mechanisms are ways in which industrial countries can redirect the responsibilities for and consequences of sustained pollution towards middle- and low-income countries (Cabello and Gilbertson, 2012). There are many examples where offsetting has had negative environmental and social impacts in the Global South. One such example is in the mountainous river valley of Uttaranchal, India, where run-of-the-river dam projects partly financed through selling carbon credits to Northern polluters have damaged local low-carbon irrigation systems (Lohmann, 2010; Gilbertson, 2008).

One of the problematic issues with regard to carbon offsetting is the delicate tightrope that many of these countries walk between promoting projects for forest carbon offsets on the one hand and vested interests that would undermine any such projects on the other. The conflict between protecting forests and promoting other industries such as the palm oil, mining and timber industries could at least in theory have created a difficult and ambiguous policy space for carbon trading, but often it has meant a rather liberal interpretation of protecting forests, often at the expense of biodiversity.

'Fragile states' are often not in a position to make the decision to forgo income from destructive industries within the largely unchanged political economy of global markets for minerals, timber and palm oil (Pearse and Böhm, 2014). Offsetting carbon for forestry/deforestation has been deeply contested as it reinforces and sometimes intensifies conflicts around land tenure for forest and land traditionally inhabited by Indigenous communities. The term 'green grabbing' is now being used to convey the way in which carbon offsets and other forms of marketised conservation revalue land and resources in ways that can alienate local populations (Fairhead *et al.*, 2012). In some cases, there is an argument that carbon markets follow a colonial path in the appropriation and destruction of land (Bachram, 2004). The shift towards using global market actors has not changed this in a dramatic way. There is an overall danger not just of green grabbing but also that many carbon markets function in a fraudulent way.

There is a danger that carbon markets in general and offsetting in particular could lead to marginalised groups being disadvantaged through both green grabbing and a rise in prices for commodities. There are a number of reasons why carbon offsetting could be understood as a form of neo-colonialism, as it favours polluting industries in the North. Table 5.1 sums up the key arguments.

The case of Plantar

Monocultural tree plantations are often referred to as 'planted forests', 'forest plantations' or 'forests'. All of these definitions gravely misrepresent the social and environmental impacts imposed by large-scale industrial tree farms. Eucalyptus plantations in Brazil are not forests but rather a monocrop cultivated within an intensive large-scale agro-industrial model.

Table 5.1 Environmental and social implications of carbon offsetting

Environmental implications	*Social implications*	*Social sustainability*	*Social policy*
The price of carbon can fall to a very low level and developed industrial countries can buy carbon cheaply, resulting in more 'business as usual'. The commodification of nature has reached a new level by allowing developed countries to buy and sell the rights to pollute, which could lead to exploitation of developing and low-income countries where emissions are cheaper, green grabbing and a continuation of business as usual.	Social justice dimensions have been neglected in the discussion around carbon markets. There is the danger that carbon markets have led to low prices for carbon and that marginalised groups in low-income countries can be exploited through CDM projects. Carbon offsetting also pulls the spotlight away from the responsibilities of developed nations and the exploitative aspects of the global neo-liberal economy. There is no proof that carbon trading can support an expensive energy transition, and no evidence that it can support existing low-carbon subsistence communities.	The concept of carbon markets can be seen as a type of neo-colonialism, as it imposes a Western ideology. It also favours polluting industries in countries where emissions are cheaper. This has led to 'low-hanging fruit' in the market, resulting in poorer areas in poorer countries becoming sites for polluting activities that are not allowed in the developed industrial world.	Green commodification relies upon the belief that green growth could simultaneously reduce poverty and prevent climate change. It also relies significantly upon the need for more CO_2 emissions for the market to function and neglects other aspects of environmental degradation such as losing intact eco-systems and a diminishing biodiversity.

However, the definition of 'planted forests' provides 'green' cover for the corporations involved and allows for the continued misclassification of tree plantations. This chapter aims to discuss the social and environmental impacts of eucalyptus plantations in the state of Minas Gerais, Brazil, operated by the local charcoal and pig-iron corporation, Grupo Plantar (hereafter referred to simply as Plantar).

Plantar began exploring access to the CDM market some 15 years ago to finance its activities. It was one of the first companies to be supported by the World Bank's Prototype Carbon Fund (PCF, now the WB Carbon Fund), which anticipated the purchase of over 1.5 million CERs (about US$25 million, assuming credits are sold at US$15) in 'emissions reductions' by 2012 (Gilbertson and Reyes, 2009).

Much has changed since those numbers were projected. Through a intensive campaign to enter the CDM market, the social consequences for the local communities, the crash of the CDM market to under $0.30 per credit and finally petitioning for voluntary cancellation of the credits, the case of Plantar depicts the inherent flaws in the carbon market and its social and environmental consequences. This section aims to probe the historical conditions and current contradictions inherent in carbon markets through the case study of Plantar.

Plantar was one of the first companies to take on the opportunity that international carbon trading offered in terms of economic return and an improvement of the enterprise's image. It applied for the registration of three distinct projects as CDM projects, each corresponding to a different stage necessary for metallurgical production. The projects have been implemented at the expense of social and environmental impacts that have engendered a knock-on effect, starting with water stress and contamination of remaining water sources, leading to hazardous chemicals having negative impacts on human and non-human life in the area. On-site workers have been most affected by health impacts and precarious work. In addition, the displacement of local communities and an increasingly repressive atmosphere has led to social divisions.

The 23,000 hectares of eucalyptus monoculture is defined by the UNFCCC as a 'forest', a key issue for monoculture tree plantations worldwide. This case study, presented as a reforestation initiative, allows us to question the rationale of carbon offsetting and to discuss the numerous underlying criticisms and injustices that are rarely considered at any length, and aims to put forward valid arguments for the exclusion of Plantar from the CDM.

Substantive dimension

Eucalyptus plantations have strong negative impacts on the basic needs of local communities, creating a sort of domino effect, which can be seen as the expression of the interdependence existing between the environment and the people. Eucalyptus is an exogenous fast-growing tree, and eucalyptus plantations cause water stress in affected territories. The impact of water stress affects local streams and waterways upon which local communities and biodiversity depend. The native *cerrado* (savannah biome) is host to small streams, but as a result of the large-scale plantations many water springs have dried up. In addition, Plantar uses several pesticides to kill competing plants and animals, causing soil depletion and polluting land and water.

Limited water supplies negatively impact basic needs, including food sovereignty. In addition, the intensive use of pesticides poisons the streams and sources of drinkable water, causing local residents give up their homes and move to urban areas. Local production has been crushed by the large-scale industry. The *cerrado* was previously used for crop production, cattle, wild foods, medicines and crafts. Local food production, agriculture and stock-raising efforts depending upon biodiversity in the area have been affected, leaving many unemployed (Lohmann, 2006: 308). As a result, the local economy has shifted to a wage-based economy, forcing more people to work for the company or leave the region.

Furthermore, the company plants one non-native tree species in order to fell and burn the trees in a seven-year rotation cycle. There is no evidence supporting carbon neutrality claims from a short, rapid-growth life cycle (Pearce, 2002). In fact, research shows that plantations do not even begin to balance the CO_2 lost from vegetation clearance and soil disruption until anywhere from ten to 100 years in an established ecosystem. Large amounts of pesticides are used in the plantations to eradicate other species that might be present. Brick ovens burn the trees into charcoal for use in the company's pig-iron operations. Forests and pastures have been destroyed to make way for the eucalyptus plantations, in the process releasing CO_2 locked in the soil. At the other end of the process lie further pollutants from the pig-iron factories. More broadly, the project contributes emissions from the entire production chain that encompasses iron smelting, shipping and so on.

All along the production chain, workers are of course the first to be exposed to the chemical products used in the plantation process or during the burning phase. Despite the company's claims of job creation, the Curvelo Regional Labour Office (DRT) contested Plantar's working conditions and issued a summons to the company in 2002 for slave and child labour in timber extraction and charcoal production. The DRT fined Plantar after finding 194 workers without any registration in its plantation near Curvelo (Lohmann, 2006: 315). The claims were so grave that the state of Minas Gerais began to investigate the company's actions. The company financed illegal outsourcing of labour in 2002, something that the Minas Gerais Parliamentary Investigation Commission subsequently confirmed. As mentioned in the official report, Plantar put workers in "precarious labour relations" and "abominable working conditions" and promoted "slave and child labour and deforestation of the *cerrado*" (Lohmann, 2006: 314). Problems with hygiene, housing, drinking water, food and transport have also been mentioned, which led the Commission to note the infringement of International Labour Organisation (ILO) provisions. Moreover, charcoal workers are constantly exposed to pesticides and are at a high risk of accidents.

As a result, the Plantar website now describes projects for enhanced worker safety, claiming that "Plantar developed the Sistema de Gestão de Segurança e Saúde Ocupacional – SGSSO [Safety and Occupational Health Management System], recognizing that this is the set of best practices, which can also contribute to be the benchmark regarding work safety and occupational health issues" (Grupo Plantar, n.d.). However, even with this "set of best practices" in place, research in 2012 found that several Plantar workers complained about long hours, dismissal without explanation or pay, severe negative health impacts and terrible working conditions (Carbon Trade Watch and FASE-ES, 2013). Furthermore, workers cannot easily organise themselves because the company creates built-in worker rotations over long distances and frequently changes worker rotations. A former Plantar worker stated: "these [Plantar] people don't want unions. They immediately co-opt the union leaders and they begin to make them a part of their inner circle of managers and directors" (Lohmann, 2003: 308).

Plantar claims to provide important jobs for the region but this is questionable because the actual number of employees is very difficult to estimate. According to Plantar's website, 1,200 direct jobs have been created in the monocultural plantations, but it does not in fact specify the duration for which each worker is hired, nor the hours worked. Furthermore, most of the work in the eucalyptus plantation is only required during the first two years, at the plantation phase, which consists of land preparation, planting, pesticide application and irrigation. Far less labour is required through the remaining four years of the growth cycle (World Rainforest Movement, 2010).

Procedural dimension

As defined in Article 12.5 of the Kyoto Protocol, emissions reductions under the CDM have to be real and measurable, to result in long-term benefits and also to be "additional to any that would occur in the absence of the certified project activity" (UNFCCC, 2007). Additionality is a necessary prerequisite for a project's eligibility and validation under the CDM. A project is considered additional "if anthropogenic emissions of greenhouse gases by sources are reduced below those that would have occurred in the absence of the registered CDM project activity" (UNFCCC, 2005: Par. 43). In other words, emissions from a CDM project should theoretically be lower than emissions that would have happened anyway.

Table 5.2 The Plantar reforestation project in Minas Gerais, Brazil

Substantive dimension	*Procedural dimension*	*Social policy dimension*
• Basic needs such as water and security were not guaranteed after the project. The lack of water supply has forced people to leave the villages where they lived. • Some local citizens (notably opponents of the plantations) have received telephone threats. • Many on-site workers have had health problems due to the toxic chemicals used during the initial disinfestation procedure (e.g. intoxications by Mirex). • No preventative measures have been carried out by the company among the communities in view of the chemical treatment of the fields. • Social cohesion has been damaged as the 'for or against Plantar' issue divided the different communities affected. • Human rights and cultural diversity have been damaged (e.g. Quilombolos and Indigenous peoples in the Espirito Santo region where Plantar manages field operations). Livelihoods have also been damaged. As shown in the documentary *The Carbon Connection* (Fenceline Films, 2007), jobs have been offered to local communities but clear data on the number of persons employed in the monoculture cultivation (and the terms of their employment) are missing.	• Most of the affected communities only discovered that the tree plantation project was happening after the environmental impacts had become apparent, primarily through water loss and pesticide pollution. • Residents were not involved in consultation processes before the project started. • Some educational activities have been organised by the company since the project has been established. There is no clear information about the real impact of these activities, nor about their frequency or the identity of the people at whom they are aimed. • Opponents have been threatened in different manners and dissuaded from continuing with their resistance. • NGOs and social movements supporting the local communities in resistance against the Plantar project organised campaigns and wrote several letters directly to the UN CDM platform and have struggled to get their comments on the web page (UNFCCC, n.d.).	Despite improvements in the quality of life for some Brazilians, resource and territorial changes illustrated the "government's political alliances with the most conservative sectors both in parliament and in the composition of ministries" (Calazans *et al.*, 2015: 106). This "resulted in weakened forest and mining codes, slashed environmental legislation, non-implementation of the territorial rights of Indigenous and traditional peoples, fishers, Quilombolas, as well as precarious and outsourced labour on petroleum rigs. These mollifications were pushed through by corporate interests, bribery and political corruption based on the game of partisan participation in the power pact from powerful elites" (Calazans *et al.*, 2015: 106).

In the CDM, additionality is assessed by a Designated Operational Entity (DOE), a private agency selected by the project participants which subsequently produces a validation report to be evaluated by the Methodology Panel of the CDM Executive Board (EB). The DOE's assessment, including baseline definition and additionality, is then presented in a Project Design Document (PDD) (UNFCCC, 2005: Par. 37). Reliance upon private agencies paid by the companies to furnish their evaluation studies raises doubts about the impartiality of the DOEs.

The CDM's additionality verification and guarantee process is ambiguous not only at the international level (the activities of the EB and DOEs) but also at the national level, where Designated National Authorities oversee CDM proposals. At this level, Axel and Katharina Michaelowa have stated, "CDM authorities do not care about additionality of CDM projects" (cited in Lohmann, 2006: 178).

The large number of CDM projects whose additionality has been questioned and contested led the EB to strengthen the verification process and provide a systematic assessment in 2007 (Gilbertson and Reyes, 2009; Michaelowa and Purohit, 2007). An increase in rejections and reviews of CDM projects resulted and by 2011 new guidelines for the establishment of performance standards were implemented (Hayashi and Michaelowa, 2013: 192). However, by 2011, many CDM projects were already coming to the end of their first term. Although several projects were reviewed, no projects that were already underway were thrown out.

The four main tools defined by the EB and used by DOEs in order to assess additionality are:

1 *positive list analysis*, affecting a specific category of CDM projects (notably those which apply to the AM0001 methodology);
2 *barrier analysis*, to prove that the CDM methodology is then seen as the only possibility that would overcome the frequent barriers that supposed low-emissions projects face;
3 *investment analysis*, to demonstrate that the CDM project is economically less advantageous than a plausible concurrent project;
4 *common practice analysis*, applied by the DOEs to prove the originality of the project and the positive counter-trend it represents compared with the ordinary scenario (Schneider, 2009: 244).

These accounting mechanisms built into the CDM process demonstrate that consultants cannot establish in a verifiable way what would have happened if the projects had never occurred. Furthermore, the bureaucracy inherent in the CDM guarantees that only the most established and connected corporate elite have access to CDM funding.

Plantar (including Plantar Carbon, Plantar Siderurgica, Plantar SA, and so forth) engaged two technical service organisations as DOEs, Tüd Süd for the reforestation methodology and Det Norvske Veritas (DNV) for the two others: methane mitigation, which produced validation reports between 2007 and 2011, and green pig-iron production. In all three cases, barrier analysis and common practice analysis were selected as the additionality assessment tools (Grupo Plantar, 2007: 17, 2009: 52, 2012: 31). They mainly underlined the presence of investment barriers, as well as management, institutional, technical and knowledge barriers. Barrier analysis is generally the tool most exploited by corporations and at the same time the tool most often defined by experts as lacking credibility due to frequent poor documentation and characterised by companies' discretionary measurements (Schneider, 2009: 244).

Discretion and incongruence are key elements in the Plantar additionality case, especially regarding validation of the Plantar reforestation project and the CER issuance that it produced. In fact, the first project submitted by Plantar in 2000 – and backed by the World Bank – was a poster-child of the CDM due to its claimed 'replicability' potential and was initially presented to the EB as a carbon sequestration project through reforestation using eucalyptus plantations. However, Plantar needed to ensure that it would have sufficient funds to replant the monocultures after the first seven-year harvest. The original reforestation project proposal was rejected by the EB. The company justified the rejection of the reforestation project as the result of UNFCCC internal regulations on land use, land use change and forestry that were still incomplete at the time of the submission (Grupo Plantar, 2009: 51).

Although Plantar claimed that there would be an "accelerated reduction in the plantation forestry base in the state of Minas Gerais" (Lohmann, 2006: 303), the company presented the 23,100-hectare eucalyptus plantation as if it were a native forest. The project stated that after harvesting the trees for pig-iron manufacture, it would not be able to replant them unless CDM finance was forthcoming. At the time Plantar owned rural properties covering more than 180,000 hectares in Minas Gerais, mainly devoted to growing eucalyptus trees for charcoal, in addition to managing lands for other eucalyptus companies covering around 600,000 hectares across Brazil.

The rejection of this project should have been enough to keep Plantar out of the CDM. However, the second time Plantar presented the project, it argued that the 23,100 hectares would provide trees to be burned in ovens to make charcoal, which would take the place of mined coal. Thus it was now presented as a 'fuel-switching' project. Again, the EB rejected this proposal, as Plantar was already producing charcoal from eucalyptus trees on its land, notably on the 180,000 hectares in Minas Gerais, which clearly demonstrated the project's lack of additionality for a second time. Even after two EB rejections, a growing social resistance movement and international opposition due to the disastrous social and environmental impacts, the project was presented for a third time, but divided into three different CDM projects, each one related to a different stage of pig-iron production.

In 2007, Plantar first gained access to the CDM for a methane reduction project, generating 112,689 CERs over a seven-year period from 2004 to 2011 (Grupo Plantar, 2007: 35). This CDM project sold credits on the carbon market to polluters in the Global North while at the same time did nothing to reduce emissions or destruction caused by eucalyptus plantations in Brazil. The simple project regulated the temperature in the ovens and tracked ventilation. The process was dressed up in technical jargon with reference to a study conducted at the Federal University of Minas Gerais (IPEAD) (Grupo Plantar, 2007: 2).

Local communities were not consulted on the CDM process. As more land was earmarked for eucalyptus production, local residents began to resist. Several letters were sent to the EB throughout the process and by 2009 a follow-up letter was sent to the EB condemning Plantar for its actions. A coalition of numerous organisations and social movements wrote:

> As far as we are concerned, Plantar SA's large-scale, chemical-intensive plantations of fast-growing eucalyptus trees and their subsequent burning can in no way be considered a mechanism for climate justice. On the contrary: the contamination and disappearance of rivers and streams; the forced displacement of peasant farmers, Indigenous forest-dwelling communities and *geraiszeiros* [inhabitants of the Cerrado Savannah ecosystem]; the land disputes over agrarian reform measures and with Quilombola [Afro-Brazilian] communities fighting to recover their ancestral territory [as is currently the case in Minas Gerais and Espírito Santo]; the destruction of native forest in the Cerrado and Atlantic Forest regions and its replacement with plantations of a single, exotic tree species; the repression, criminalization and intimidation of local community leaders and resistance movements; the threat to food security in areas around eucalyptus plantations; outsourcing, precarious work conditions and high rates of work-related accidents and disease [as amply documented by many sources] – all of these are essential elements that should be taken into consideration and lead the CDM Executive Board to reject Plantar SA's project proposal once again.
>
> (Coalition Against Plantar Reforestation Project, 2013)

After a multitude of international efforts to keep Plantar out of the CDM, in 2010 the Plantar project was finally registered and issued just over four million CERs in association with the World Bank's Prototype Carbon Fund (PCF) and the BioCarbon Fund (BioCF) for three project activities referred to as *Projeto Plantar* (Plantar Project) and managed by Plantar Carbon Ltda (Grupo Plantar, n.d.). The Plantar Project was then divided into three connected CDM projects: the methane avoidance project, which was registered by the EB in 2007; the reforestation project, finally registered in 2010; and the green pig-iron production, registered in 2012.

The three projects include the reforestation methodology, which is now referred to as "establishing new stocks of planted forests", using terminology like "new stocks" to prove additionality and "planted forests" to cloak the word "plantations". The "Reforestation as Renewable Source of Wood Supplies for Industrial Use in Brazil" project was accepted by the EB in 2010, although there were nine letters of opposition sent by civil society organisations (UNFCCC, n.d.). The project will generate an estimated 75,783 metric tonnes of annual CO_2e reductions per year.

The second project is a continuation of the methane project, registered in 2007 and concluded in 2011, for "improving the production process of renewable charcoal in order to reduce methane (CH_4) emissions" (Grupo Plantar, 2007: 2). The third project, registered in 2012 with a US$12.3 million commitment from the World Bank, is for the "production of Green Pig Iron®, based on the use of renewable charcoal, instead of coal coke" (Grupo Plantar, 2012; CDM Monitoring Report ref. n 7577, 2015: 1). The CDM certification of the three projects will run for 30 years. The World Bank immediately purchased a share of the CERs through the PCF and BioCF in a structured financial transaction with Rabobank International.

These repeated stages of rejection and modification show modification both in nomenclature and analysis, without constituting any change in the actual project. The reforestation project was rejected twice before 2010 and perceived as neither additional nor sustainable by the EB. However, the third attempt was accepted, which included backdating the counting period for tCERs (the period of eligibility for temporary carbon credits specific for the forestry sector) to November 2000 (Grupo Plantar, 2009: 25). The fact that the reforestation project was finally registered in 2010 after a ten-year power struggle between the World Bank, the EB, civil society organisations and Plantar, and allowed the backdating of the project to generate credits from 2000, should negate any additionality argument that the EB, the UN, the World Bank or Plantar can claim.

By 2012, the carbon market was bottoming out and the CDM in particular was looking like it might not survive. The EB built in an escape clause to cancel what it considered an over-supply of credits, driving down the price while the CDM was "faced with possible collapse because demand in recent years from the principal buyers – countries tasked with emission reduction obligations under the Kyoto Protocol – has dropped" (Wambi, 2014). This escape clause made it possible for companies with CDM credits to voluntarily cancel them from the UNFCCC register, thereby subtracting CO_2 from both the atmosphere and carbon markets, in a demonstration of engagement with socio-environmental and climate protection issues. Despite years of angling to enter the CDM under the premise that it would not be able to function without the CER revenue (and finally entering with three projects), Plantar voluntarily cancelled 278,116 tCERs between 2012 and 2014, permitting various investors to take credit for hypothetical environmentally responsible actions. This should not be seen as anything more than a smokescreen, however, given the environmental and social impacts of Plantar's operations in the region. Despite the validation obtained from the private agencies and the EB, the sustainability of this project is highly questionable. Marcelo Calazans of FASE-ES puts it like this:

> The Plantar Project should never be seen as a sustainable project, as the [growth] cycles are very short, it is large-scale, the quantity of toxic chemicals is immense and has poisoned many workers. We interviewed a lot of workers who have suffered from MIREX [ant insecticide] contamination. I always ask these companies how much they invest in the cultivation, and within this sum how much is for security services. You can be sure that the majority of the employees in the fields are armed security guards working to ensure the property's security.
>
> (Calazans, 2015)

With the use of private security forces, there is no surprise that a climate of insecurity is established in the monocultural areas. The award-winning film *The Carbon Connection* (Fenceline Films, 2007) documented how local communities were exploited by Plantar for the 12,540 hectares needed for its World Bank PCF reforestation project proposed in 2003. At the time of filming, in 2005, members of the community came together to speak out against the company and the impacts that the plantations were having on their lives. One tactic that the company used was to offer jobs to local people who had the courage to speak out, or to their families. Certain communities came together to organise against Plantar's atrocious practices, but were silenced by a consistent pattern of manipulation and intimidation, including anonymous phone calls that threatened that 'accidents' could occur, more pointed threats on people's lives and even death threats aimed at other family members (Fenceline Films, 2007). All of these elements have had an important effect on social cohesion in local communities, forcing inhabitants to take a position on Plantar in the region. The constant threat of Plantar creates divisions among locals who claim to be in favour or against the company, advancing general mistrust.

Following the threats to local citizens, several other organisations began a campaign to keep Plantar out of the CDM. Although the letter-writing and other campaigns caused an international uproar, with the help of the World Bank Plantar still managed to enter the CDM. Plantar continues to use tactics that silence its opposition and instead of being criminalised, it is hailed by the World Bank, the UN and the Forest Stewardship Council as a "green" and innovative business (Grupo Plantar, n.d.).

The steps taken by Plantar and the World Bank underscore the lack of credibility of the CDM and emphasise the powerful role that businesses and institutions have in determining and defining 'sustainable development'. The Plantar Project is just one of many projects in the CDM that has little positive impact on the environment. The Plantar/World Bank Project represents a fallacy embedded in the CDM including the entire concepts of "additionality", determining baselines and offset methodologies in general (Lund, 2010: 278). In the case of Plantar, the ambiguity of additionality rules, the power relations, the backdating of credits, the third-party private verifiers and the cancellation of credits are all symptoms of the greater flaws inherent in a system that involves the trading of fictitious commodities such as pollution.

Social implications, challenges and solutions

This chapter has explored the social implications of the Plantar Project and the social and environmental harm inflicted by the procedures implemented by the company, including damage to land, water, human health and local community livelihoods. The Plantar Project, as is the tendency with so many other carbon offset projects, followed a one-size-fits-all design that did not deal with the real complexities and intricacies of communities and livelihoods on the one hand and biodiversity and intact ecosystems on the other. Scholars have pointed out that extractive industries often lead to the economic and social marginalisation of residents in remote and rural areas. Often local people obtain few of the benefits that come with mining and other extractive activities on their land. It is more likely that these activities will threaten existing and viable livelihoods (Cademartori, 2002; Freudenburg, 1992).

As mentioned earlier, it is hard to evaluate the positive impact of monocultural plantations on employment, as extensive work is only required during the first two years, at the plantation phase, which consists of land preparation, planting, pesticide application and irrigation. Far less labour is required through the remaining four years of the growth cycle (World Rainforest Movement, 2010). Of equal importance is the negative impact that these activities have on the environment. As stated earlier, it is questionable whether these projects contribute to reducing CO_2 emissions. A monocultural plantation threatens the local biodiversity, which has a great impact on ecosystems. This chapter supports a large body of evidence that demonstrates how low-carbon development – as understood through some offsetting projects such as the Plantar Project – contributes to a fraudulent understanding of climate change mitigation and neglects many social and environmental implications that go beyond greenhouse gas emissions.

It is therefore important to develop sustainability and environmental justice beyond a traditional human-centric discourse. Scholars such as John Drake and Reuben Keller (2004) and Mick Hillman (2006) have explored the importance of ecological integrity. There is a good argument that we need to have a change in our global understanding of development and that resource-intensive development threatens the environment through climate change and the biodiversity and ecological integrity of

Table 5.3 Social implications, challenges and solutions

Social implications	*Challenges*	*Solutions*
Afforestation and reforestation projects are a very questionable typology of CDM. They represent a minority within the CDMs as only 56 are actually registered (out of 7,668). Mainly of large scale and carried out by major corporations, these projects are frequently an opportunity for the eucalyptus cultivation sector to expand under the pretext of promoting sustainable development. The water stress caused by these exogenous trees triggers a knock-on effect to the detriment of local communities, who are forced in several cases to leave their land in urban flight.	Local communities, NGOs and civil society organisations have complained and reached out to the EB, but with little or no response and positive results for Plantar. Removing Plantar from the CDM does not even seem to be an option. There is a power accord between the World Bank, the EB and Plantar. Carbon trading and the CDM, as neo-liberal strategies, have become the dominant frames for understanding climate change mitigation, but they prevent the development of more effective and socially just solutions and threaten the biodiversity of several ecosystems including the *cerrado* and the *Mata Atlantica* forest, to name a few.	There is a need to rethink CDM projects such as Plantar's as they are neither environmentally or socially sustainable and it is questionable whether they are mitigating climate change. Strict regulations should be enforced on CDM projects so that projects such as Plantar's are not credited. Support should be given to local communities, workers and citizens who are struggling against the negative impacts of eucalyptus plantations in Brazil and worldwide. The EB should be held accountable for human and environmental harms caused by the companies whose projects they approve. The flaws inherent in carbon trading should be addressed and the focus should be shifted to keeping fossil fuels underground.

many habitats. There is a real threat that our ecosystems will not function and their integrity will be compromised to the extent that they will have difficulty in supporting life (Pimentel *et al.*, 2000; Earth Charter Initiative, 2010). Ecological integrity is important as it highlights the inherent potential, stability, capacity for self-repair and independent management of ecosystems (Karr, 1992). It is these features that enable ecosystems to provide, regulate and support all life (Millennium Ecosystem Assessment, 2005). Indigenous peoples have often upheld the important notion that humans are inseparable from other living things and see the environment as an interconnected community (LaDuke, 1999; McGregor, 2009).

The ecological integrity approach has resulted in an attempt to incorporate environmental concerns into a movement that has often been perceived to be anthropocentric (Shrader-Frechette, 2002):

> When we interrupt, corrupt, or defile the potential functioning of ecological support systems, we do an injustice not only to human beings, but also to all of those non-humans that depend on the integrity of the system for their own functioning.
>
> (Schlosberg, 2013: 44)

This is an important development, as it looks at the functioning and capabilities of individuals but also communities and the environment that forms the land of those communities. There is a threat that monocultural plantations, like the type cultivated by Plantar, will continue to expand and threaten both biodiversity and the communities and Indigenous people who depend upon the environment. It is therefore important that we rethink low-carbon development and give consideration to the other important environmental aspects that, if neglected, might have tremendous negative impacts on both the environment and the communities that depend upon it.

References

Bachram, H. (2004), "Climate fraud and carbon colonialism: the new trade in greenhouse gases", *Capitalism Nature Socialism*, 15(4), 5–20.

Cademartori, J. (2002), "Impacts of foreign investment on sustainable development in a Chilean mining region", *Natural Resources Forum*, 26(1), 27–44.

Cabello, J. and Gilbertson, T. (2012), "A colonial mechanism to enclose lands: a critical review of two REDD+-focused special issues", *Ephemera*, 12(1–2), 162–80.

Calazans, M. (2015), Interview with the author, August.

Calazans, M., Gilbertson, T. and Meirelles, D. (2015), "Not one more well!: corruption and Brazil's pre-salt expansion", in Temper, L. and Gilbertson, T. (eds), *Refocusing Resistance to Climate Justice: COPing In, COPing Out and Beyond Paris*, EJOLT Report No. 23, pp. 104–9, available at: www.ejolt.org/wordpress/wp-content/uploads/2015/09/climate-justice-report.pdf (accessed 29 June 2016).

Carbon Trade Watch and FASE-ES (2013), *Like Oil and Water: Struggles Against the Brazilian Green Economy*, available at: www.carbontradewatch.org/multi-media/en/like-oil-and-water-struggles-against-the-brazilian-green-economy (accessed 8 July 2016).

CDM Monitoring Report ref. n 7577 (2015), "CDM Monitoring Report ref. n 7577".

Coalition Against Plantar Reforestation Project (2013), "Letter to the Executive Board – Plantar SA CDM project: global warming continues unabated", available at: http://carbonmarketwatch.org/wp-content/uploads/2010/03/plantar_letter-to-the-executive-board.pdf (accessed 23 March 2016).

Daly, H. (1996), *Beyond Growth: The Economics of Sustainable Development*, Boston, MA: Beacon Press.

Daly, H. (2015), "Commentary: how best to value nature", *Heinrich Böll Stiftung*, 6 November, available at: www.boell.de/en/2015/11/06/comment-jutta-kill-herman-daly (accessed 23 March 2016).

Drake, J. M. and Keller, R. P. (2004), "Environmental justice alert: do developing nations bear the burden of risk for invasive species?", *Bioscience*, 54(8), 718–19.

Earth Charter Initiative (2010), *The Earth Charter*, San Jose: ECI, available at: http://earthcharter.org/discover/download-the-charter/ (accessed 23 March 2016).

Fairhead, J., Leach, M. and Scoones, I. (2012), "Green grabbing: a new appropriation of nature?", *Journal of Peasant Studies*, 39(2), 237–61.

Fenceline Films (2007), *The Carbon Connection.*

Freudenburg, W. R. (1992), "Vulnerable localities in a changing world economy", *Rural Sociology*, 57(3), 305–32.

Gilbertson, T. (2008), "Bhilangana Dam on troubled waters", *Mausam: Talking Climate in Public Space*, 1–2(2–5), 22–3.

Gilbertson, T. and Reyes, O. (2009), *Carbon Trading: How It Works and Why It Fails*, Uppsala: Dag Hammarskjöld Foundation.

Grupo Plantar (n.d.), website, available at: www.grupoplantar.com.br/ (accessed 23 March 2016).

Grupo Plantar (2007), "Project Design Document: Mitigation of Methane Emissions in the Charcoal Production of Plantar, Brazil", Project ref. n 1051, UNFCCC.

Grupo Plantar (2009), "Project Design Document: Reforestation as Renewable Source of Wood Supplies for Industrial Use in Brazil", Project ref. n 2569, UNFCCC.

Grupo Plantar (2012), "Project Design Document: Use of Charcoal from Renewable Biomass Plantations as Reducing Agent in Pig Iron Mill in Brazil", Project ref. n 7577, UNFCCC.

Hayashi, D. and Michaelowa, A. (2013), "Standardization of baseline and additionality determination under the CDM", *Climate Policy*, 13(2), 191–209.

Hillman, M. (2006), "Situated justice in environmental decision-making: lessons from river management in southeastern Australia", *Geoforum*, 37(5), 695–707.

IEA (International Energy Agency) (2008), *World Energy Outlook 2008*, Paris: IEA.

IMF (International Monetary Fund) (2008), *World Economic Outlook*, Washington, DC: IMF.

Karr, J. R. (1992), "Ecological integrity: protecting earth's life support systems", in R. Costanza, B. Norton and B. Haskell (eds), *Ecosystem Health: New Goals for Environmental Management*, Washington, DC: Island Press, pp. 223–38.

Kollmuss, A., Schneider, L. and Zhezherin, V. (2015), *Has Joint Implementation Reduced GHG Emissions?: Lessons Learned for the Design of Carbon Market Mechanisms*, Stockholm: Stockholm Environmental Institute.

LaDuke, W. (1999), *All Our Relations: Native Struggles for Land and Life*, Cambridge, MA: South End Press.

Lohmann, L. (ed.) (2003), *Certifying the Uncertifiable: FSC Certification of Tree Plantations in Thailand and Brazil*, Montevideo: World Rainforest Movement.

Lohmann, L. (2005), "Marketing and making carbon dumps: commodification, calculation and counterfactuals in climate change mitigation", *Science as Culture*, 14(3), 203–35.

Lohmann, L. (2006), "Carbon trading: a critical conversation on climate change, privatisation and power", *Development Dialogue*, 48, 1–362.

Lohmann, L. (2010), "Neoliberalism and the calculable world: the rise of carbon trading", in K. Birch and V. Mykhnenko (eds), *The Rise and Fall of Neoliberalism: The Collapse of an Economic Order?*, London: Zed Books, pp. 77–94.

Lohmann, L. (2011), "Capital and climate change", *Development and Change*, 42(2), 649–68.

Lund, E. (2010), "Dysfunctional delegation: why the design of the CDM's supervisory system is fundamentally flawed", *Climate Policy*, 10(3), 277–88.

McGregor, D. (2009), "Honouring our relations: an Anishnaabe perspective on environmental justice", in: J. Agyeman, P. Cole, R. Haluza-Delay and P. O'Riley (eds), *Speaking for Ourselves: Environmental Justice in Canada*, Vancouver: UBC Press, pp. 27–41.

Michaelowa, A. and Purohit, P. (2007), "Additionality determination of Indian CDM projects", Discussion Paper CDM-1, Climate Strategies, available at: www.internationalrivers.org/files/attached-files/additionality-cdm-india-cs-version9-07.pdf (accessed 8 July 2016).

Millennium Ecosystem Assessment (2005), *Ecosystems and Human Well-Being: Synthesis*, Washington, DC: Island Press, available at: www.millenniumassessment.org/documents/document.356.aspx.pdf (accessed 23 March 2016).

Milner, M. (2007), "Global carbon trading market triples to $30b", *The Hindu*, 5 May.

Pearce, F. (2002), "Tree farms won't halt climate change", *New Scientist*, available at: www.newscientist.com/article/dn2958-tree-farms-wont-halt-climate-change/ (accessed 8 July 2016).

Pearse, R. and Böhm, S. (2014), "Ten reasons why carbon markets will not bring about radical emissions reduction", *Carbon Management*, 5(4), 325–37.

Pimentel, D., Westra, L. and Noss, R. (2000), *Ecological Integrity: Integrating Environment, Conservation and Health*, Washington, DC: Island Press.

Porritt, J. (2005), *Capitalism As If the World Matters*, London: Earthscan.

Reyes, O. (2011), "EU Emissions Trading System: failing at the third attempt", *Carbon Trade Watch*, available at: www.carbontradewatch.org/downloads/publications/ETS_briefing_april2011.pdf (accessed 23 March 2016).

Schneider, L. (2009), "Assessing the additionality of CDM projects: practical experiences and lessons learned", *Climate Policy*, 9(3), 242–54.

Schlosberg, D. (2013), "Theorising environmental justice: the expanding sphere of a discourse", *Environmental Politics*, 22(1), 37–55.

Scott, M. (2008), "Market meltdown? Carbon trading is just warming up", *Independent on Sunday*, 27 July.

Shrader-Frechette, K. (2002), *Environmental Justice: Creating Equality, Reclaiming Democracy*, New York: Oxford University Press.

Spash, C. L. (2010), "The brave new world of carbon trading", *New Political Economy*, 15(2), 169–95.

Stern, N. (2006), *Stern Review: The Economics of Climate Change*, London: HM Treasury, available at: http://webarchive.nationalarchives.gov.uk/20100407172811/www.hm-treasury.gov.uk/stern_review_report.htm (accessed 15 March 2016)..

Stern, N. (2007), *The Economics of Climate Change*, Cambridge: Cambridge University Press.

Stern, N. (2009), *A Blueprint for a Safer Planet: How to Manage Climate Change and Create a New Era of Progress and Prosperity*, London: Bodley Head.

UNDP (United Nations Development Programme) (2007), *Human Development Report 2007/8*, New York: Palgrave Macmillan.

UNEP (United Nations Environment Programme) (n.d. a), "CDM/JI Pipeline Overview Page", available at: www.cdmpipeline.org/overview.htm (accessed 29 June 2016).

UNEP (United Nations Environment Programme) (n.d. b), "JI Projects", available at: www.cdmpipeline.org/ji-projects.htm (accessed 29 June 2016).

UNFCCC (United Nations Framework Convention on Climate Change) (n.d.), "Submission of comments to the DOE/AE, CDM: Reforestation as Renewable Source of Wood Supplies for Industrial Use in Brazil", available at: https://cdm.unfccc.int/Projects/Validation/DB/DWXTGTQLAORUROS9KPVZJUSGI8UK70/view.html (accessed 23 March 2016).

UNFCCC (United Nations Framework Convention on Climate Change) (1997), *The Kyoto Protocol*, Bonn: UNFCCC, available at: http://unfccc.int/resource/docs/convkp/kpeng.pdf (accessed 18 October 2012).

UNFCCC (United Nations Framework Convention on Climate Change) (2005), 9th Plenary Meeting, 3/CMP.1, Annex.

UNFCCC (United Nations Framework Convention on Climate Change) (2009), *Copenhagen Accord*, Bonn: UNFCCC, available at: http://unfccc.int/resource/docs/2009/cop15/eng/11a01.pdf (accessed 18 October 2012).

UNFCCC (United Nations Framework Convention on Climate Change) (2010), *Cancun Agreements*, Bonn: UNFCCC, available at: http://unfccc.int/resource/docs/2010/cop16/eng/07a01.pdf#page=2 (accessed 18 October 2012).

UN-REDD (United Nations Collaborative Programme on Reducing Emissions from Deforestation and Forest Degradation in Developing Countries) (2010), "MRV and monitoring for REDD+ implementation", available at: www.un-redd.org/Newsletter10/MRV_and_Monitoring/tabid/4864/language/en-US (accessed 29 June 2016)

Urban, F. and Nordensvärd, J. (2013), *Low Carbon Development: Key Issues*, London: Routledge.

Wambi, M. (2014), "Q&A: why Kyoto's Clean Development Mechanism is at a crossroads", Inter Press Service, 4 December, available at: www.ipsnews.net/2014/12/qa-why-kyotos-clean-development-mechanism-is-at-a-crossroads/ (accessed 23 March 2016).

World Bank (2008), *State and Trends of the Carbon Market 2008*, Washington, DC: World Bank.

World Rainforest Movement (2010), "Brazil: once again opposing Plantar's CDM project", *World Rainforest Movement Bulletin* 151.

6 Domestic energy efficiency policy and fuel poverty in England

Carolyn Snell with Johan Nordensvärd

Low-carbon transformation and fuel poverty

Energy is a complex policy area with global, national and local environmental and social impacts and policy problems. First, in environmental terms, anthropogenic climate change is considered to be one of the greatest threats to human security (UNFCCC, 2012, 2013). The need to reduce carbon emissions has dominated the global environmental policy agenda since the 1990s, and while at the time of writing there are currently no climate targets in place (the lifespan of the Kyoto Protocol having expired), international discussions continue to seek new targets (UNFCCC, 2013). While much of the debate focuses on the role and responsibility of rapidly developing/newly industrialised countries within reduction targets, climate negotiations typically start from the view that developed countries must take some responsibility for climate change and resulting policies, given their current and historic emissions (Snell and Haq, 2014).

At the EU level, for example, a variety of mechanisms are in place (such as the EU ETS) in order to reduce carbon emissions. Equally, at the national level, in the UK, the Blair government set a legally binding target of an 80 per cent reduction in carbon emissions by 2050 (Climate Change Committee, 2015). There has been increasing policy concern about energy security, partly due to dwindling energy supplies but also as a result of geopolitics (especially where energy resources are concentrated in politically unstable regions), population growth and development (Sovacool and Mukherjee, 2011; Pollitt, 2012; Chester, 2010). While UK Coalition government policy continued to follow the UK climate targets set under the Climate Change Act, there was also a noticeable increase in discussions of energy security (Davey, 2014).

Second, in social terms, despite the unsustainable use of fossil fuels, access to energy is unequally distributed, both between and within countries (Sovacool *et al.*, 2014). At the global level, the poorest countries have typically contributed little to the problem of climate change but are often exposed to the greatest effects associated with rising sea levels, reduced crop productivity and desertification, while simultaneously not benefiting

Table 6.1 Environmental and social implications of domestic energy policy

Environmental implications	*Social implications*	*Social sustainability*	*Social policy*
Energy generated from fossil fuels is associated with localised forms of pollution in addition to climate change (Wilkinson *et al.*, 2007).	**Production of energy:** UK climate change effects include increased risk of flooding, cold periods and dry spells (DEFRA, 2012). Increased health risks are associated with proximity to power stations (Wilkinson *et al.*, 2007). **Consumption of energy:** Other than climate change, most social implications relate to the under-consumption of energy. As regards negative health outcomes, lower household temperatures are associated with increased rates of respiratory diseases, strokes and cardiovascular conditions. Excess Winter Mortality (EWM) figures are attributed in part to low indoor temperatures, damp and mould (NICE, 2015). Risks to children relate to educational outcomes and development. There are also increased risks of social exclusion and mental illness (anxiety and depression) (Hills, 2012; Marmot Review Team, 2011).	**Production of energy:** There are inequalities associated with the siting of power stations – for example, pollution injustice arguments (McCauley *et al.*, 2013). **Consumption of energy:** Certain groups – children, old people, disabled people, those with underlying health conditions – are more vulnerable to the negative health effects of fuel poverty (and are also often more likely to be poor and thus affected) (Marmot Review Team, 2011).	While policy must recognise the need for behavioural changes in domestic energy practices, it also must recognise that some households are vulnerable to fuel poverty. There are justice questions surrounding policy decisions including the distributional impact of energy efficiency and fuel poverty policies, questions of procedural justice (Walker and Day, 2012) and issues of recognition of particular socio-economic groups' needs (Snell *et al.*, 2015).

from reliable sources of energy (Sovacool and Dworkin, 2014). Even within the world's richest countries, such as the UK, large inequalities exist, with up to an additional 35,000 individuals dying prematurely during the winter months, attributed in part to their being unable afford to heat their homes sufficiently (NICE, 2014). This social dimension of energy adds a layer of complexity to the nature of the problem – described by some as a 'wicked' policy problem with inherent conflicts, tensions and contradictions (Kerley, 2012; Gupta *et al.*, 2014) – and also means that policy solutions are less straightforward. It is the tension that exists within developed countries such as the UK[1] that this chapter focuses on.

Fuel poverty and domestic energy (in)efficiency in England

Fuel poverty has been a concern in the UK since the 1990s, when the combination of low incomes, poor energy efficiency and high fuel prices was blamed for a situation in which households were unable to heat their homes sufficiently and as a result suffered the potentially fatal effects of exposure to cold or damp (Marmot Review Team, 2011), debt, or having to make stark compromises with other areas of household expenditure, such as reducing the quality and quantity of food consumed (Lambie-Mumford and Snell, 2015).

The first definition of fuel poverty used by policy-makers was proposed by Brenda Boardman and considered a household to be fuel poor if they needed to spend more than 10 per cent of their income to maintain sufficient levels of warmth (Boardman, 1991). Using this measure, the Labour government introduced a target to eradicate fuel poverty entirely by 2016. Both the measure of fuel poverty and the targets were revised by the Coalition government.[2] The new measure of fuel poverty defines a fuel-poor household to be one on a low income with high energy costs, which was a measure proposed by John Hills in 2012, and the new targets require the energy efficiency band of a fuel-poor home to be no lower than Band C (with the current poorest rating being Band G) by 2030 (DECC, 2013).

As described above, there is the potential for the environmental aspects of energy policy to be in conflict with the social dimensions; for instance, a tax on fossil fuels may reduce their use, but may also harm the poorest in society, making it impossible for them to afford to use enough energy to stay healthy. Equally, policies that only consider the social dimensions of energy – for example, those that subsidise the cost of fossil fuels – may have desirable social effects, improving quality of life, but with the negative effect of increased carbon emissions (Cahill, 2001). The majority of policy-makers and academics have concluded that these distinct dimensions of energy policy *can* be addressed simultaneously, but that the most effective policies combine both carbon reduction measures and poverty alleviation, usually through energy efficiency measures targeted towards those in the most energy-inefficient housing and on the lowest incomes

(Huby, 1998; Cahill, 2001; Fitzpatrick, 2012; Snell and Thomson, 2013). Indeed, the nature of the UK's housing stock makes domestic energy efficiency improvements a promising way of addressing both policy needs (Gupta *et al.*, 2014; Dowson *et al.*, 2012). The most recent data and policy guidance highlight the importance of household energy use, given that this accounts for more than a quarter of energy use and carbon emissions in the UK (DECC, 2013: 5).

In comparing the UK with other European countries with similar heating needs and affluence, the Association for the Conservation of Energy (ACE) finds the UK to be the worst performing on six indicators of energy efficiency and fuel poverty:

> The UK is ranked [in the lowest categories] for ... homes in poor state of repair (11 out of 15), thermal performance (6 out of 8), and the gap between current thermal performance and what the optimal level of insulation should be in each country (7 out of 8).
>
> (ACE, 2013: 1)

Unsurprisingly, the Department for Energy & Climate Change (DECC) outlines that household energy represents a substantial opportunity to reduce carbon emissions and goes on to suggest that the national carbon reduction targets will be unreachable without improvements to domestic energy efficiency (DECC, 2013: 6). The age of buildings is a significant factor since building regulations have only addressed energy conservation since 1965 and have been subject to increasingly strict amendments since then (DECC, 2013: 25). While the newest housing is typically the most energy efficient, the opposite is true of the older housing stock. Where energy efficiency measures (such as loft insulation) are in place, these are often inferior compared to modern standards, and heating systems and insulation measures are often harder to modernise (DECC, 2013: 25). In 2012, around 30 per cent of the housing stock was in the three least efficient categories.

English domestic energy efficiency policy, 2010–15

By the end of its term in office in 2010, the Labour government had implemented both state-funded energy efficiency programmes and private-sector initiatives. The most notable state programme was the Warm Front Scheme, which operated at the English policy level and provided energy efficiency improvements to households on income-related benefits. Private schemes were funded through energy companies, the Carbon Emissions Reduction Target (CERT) supported energy efficiency improvements such as cavity wall and loft insulation, and much of the fund was open to all households, regardless of income, whereas the Community Energy Savings Programme (CESP) was area based and focused on energy efficiency improvements in areas of socio-economic deprivation in England, Scotland and Wales. However, the Conservative–Liberal Coalition government focused on a market-based approach to energy policy generally, and to domestic energy efficiency specifically (Hamilton *et al.*, 2014; Mallaburn and Eyre, 2014).

Political rhetoric emphasised the need to reduce state spending, and policy shifted the responsibility for the funding and delivery of most measures from the state to the private sector. By the time of the General Election in May 2015, no substantive state-funded programmes remained, with responsibility for programmes being placed in the hands of energy suppliers and consumers, and changes also being made to CERT and CESP.[3] Overall, given the budget cuts, Coalition policy has resulted in substantially less funding for domestic energy efficiency schemes (Hills, 2012).

The six main schemes relating to domestic energy efficiency in England are outlined in Table 6.2. While the Big Energy Saving Network and Fuel Poverty and Health Booster Fund are both publicly funded, these have very small budgets of around £1 million (and the former is an advice service). The Central Heating Fund (CHF) has a more generous budget of £25 million; however, at the time of writing, it is unclear how funds are being spent or distributed. The transition to smart meters is a well-funded scheme (at an estimated cost of £350 per bill over the roll-out period (House of Commons Committee of Public Accounts, 2011); however, its focus is on the provision of information and behavioural change – an assumed energy saving of 3 per cent (House of Commons, 2014) – rather than of physical improvements to the housing stock. Of the schemes identified in Table 6.2, only the Green Deal and the Energy Company Obligation (ECO) provide a substantial programme of energy efficiency measures to domestic households, and it is these schemes that this chapter now focuses on.

Case Study 1: the Green Deal

Launched in 2013, the Green Deal was the Coalition government's flagship energy efficiency programme. It aimed to encourage households to undertake energy-saving measures (such as cavity wall insulation and double glazing) through a loan paid back through energy bills, with an interest rate of around 7 per cent. The Green Deal was made available to owner-occupiers and those in the social and private rented sector with permission required from both landlords and tenants, with either being able to apply for a loan. The loan or associated debt was based on the property rather than the household, and had to be declared at the point of sale or rental.

The Green Deal excluded a variety of households (and a second scheme, the ECO, was put in place to address these). Households were excluded because "The Green Deal's Golden Rule states that the energy savings a property makes ... must be equal to or more than the cost of implementing

Table 6.2 Domestic energy efficiency policies under the Coalition government

Scheme	*Description*
Smart meters	The Coalition government set a target for all homes and small businesses to have smart meters by 2020. This has in part been led by EU policy, although UK policy goes beyond EU requirements. The roll-out will involve "replacing over 53 million gas and electricity meters. This will involve visits to 30 million homes and small businesses" (DECC, 2015a: Appendix 7). Smart meters are said to have two functions: to enable households to have a clearer understanding of the cost of their energy use on a daily basis (i.e. cost) and to encourage energy efficiency (e.g. behavioural change) (Energy Saving Trust, 2015).
Green Deal	The Green Deal allows householders and businesses to make energy efficiency improvements with some or all of the cost paid for from the savings on their energy bills. The Green Deal is intended to encourage householders to improve the energy efficiency of their homes, and only caters for properties where the energy efficiency can be improved relatively easily. It is explicitly not targeted at householders likely to be in fuel poverty, whose needs are addressed instead through ECO.
Energy Company Obligation (ECO)	ECO is a subsidy from energy suppliers that works alongside the Green Deal to provide energy-saving home improvements to those who are unlikely to benefit from the Green Deal (for instance, householders on low incomes or those in hard-to-treat properties).
Central Heating Fund (CHF)	The CHF was launched on 26 March 2015, and is a £25 million capital funding programme aimed at supporting approximately 8,000 households considered to be experiencing some of the most severe levels of fuel poverty (DECC, 2015d: 4–6). The scheme provided funding to local authorities through a competitive bidding process that ran between March and June 2015. The purpose of the CHF is described by DECC as being "to incentivise the installation of first-time central heating systems in fuel-poor households who do not use mains gas as their primary heating fuel" (DECC, 2015d: 6). The scheme aims to help to meet the new statutory target of raising energy efficiency ratings of fuel-poor homes to Band C by 2030 (DECC, 2015d: 6). At the time of writing, it is unclear which local authorities have successfully received funding. The value of this scheme should be highlighted at this juncture given that it is small compared to the value of ECO or the Winter Fuel Payments (WFP)s.

continued

Table 6.2 Continued

Scheme	*Description*
Fuel Poverty and Health Booster Fund	In 2015, DECC provided a fund of £1 million to support health- and fuel-poverty-related initiatives. The Fuel Poverty and Health Booster Fund funded projects in nine local councils. Schemes include training for health and social care professionals in order to help to identify and support fuel-poor households in Durham, and energy efficiency measures in Dudley (DECC, 2015e).
Big Energy Saving Network	The BESN is an initiative launched in 2013 with a budget of £1 million per year. The aim of the programme is to develop a network of voluntary and community-sector organisations that could work with a broader range of vulnerable customers than typically reached by national or local programmes. In its first year (2013/14), the network provided grants to 94 organisations which then worked with vulnerable consumers and provided information about energy markets, tariffs, switching and access to schemes such as ECO (DECC, 2015c). Given the 'stickiness' of vulnerable customers (in other words, they are less likely to switch or shop around) and the proximity of voluntary and community-sector organisations to vulnerable customers, the aim of the scheme is to provide 'trusted intermediaries' who can support the specific needs of vulnerable households.

the changes in the first place" (Green Deal Initiative, 2015). Typically households that were deemed 'hard to treat' (for example, because they have solid walls that cannot be insulated easily) did not meet the Golden Rule. Households at risk of fuel poverty were also unlikely to meet the Golden Rule as they often have low energy expenditure and as such "are unable to meet the savings required without subsidy" (Duxbury, 2012).

Amid concerns of low take-up among consumers, incentives were added in 2014, including the Home Improvement Fund, which offered up to £7,600 cashback for work undertaken (Ellis, 2014), with £120 million in funding offered in June 2014, a further £30 million in December 2014 and another £70 million in March 2015. However, in July 2015 the DECC withdrew funding from the Green Deal Financing Company, meaning that no new loans would be made (Vaughan, 2015; BBC, 2015).

Since it was announced in 2010, there has been substantial research into the Green Deal's effectiveness (Guertler, 2012; Snell and Thomson, 2013; Consumer Focus, 2011; Dowson *et al.*, 2012; ACE, 2012). Criticisms have typically focused on its appeal for consumers and industry, and whether it will be able to deliver the carbon reductions needed in order to meet national targets (Dowson *et al.*, 2012; Mallaburn and Eyre, 2014). In the programme's first year of operation, the government anticipated 10,000 households signing up for a Green Deal; however, by September 2014 only 4,000 were in place (the initial policy rhetoric referred to 'millions' of households and businesses benefiting from the energy efficiency improvements (DECC, 2010)). In its September 2014 report on the Green Deal, the Select Committee on Energy & Climate Change found that it had had a limited effect in terms of improving domestic energy efficiency:

> at the end of 2013, 4.5 million cavity walls and 7 million solid walls remained uninsulated, with a carbon saving potential of 2 $MtCO_2$ and 5 $MtCO_2$, respectively. An additional 0.7 $MtCO_2$ could be saved as a result of addressing 9 million lofts that still required additional insulation filling.... The Government's latest statistics reveal that estimated lifetime carbon savings through Green Deal plans have so far been negligible, at 0.04 $MtCO_2$ through Green Deal plans and 0.07 $MtCO_2$ through Green Deal cashback.
>
> (Select Committee on Energy & Climate Change, 2014)

In addition to this, the Green Deal has been criticised for not supporting decarbonisation measures (Snell and Thomson, 2013).

In environmental terms, then, the Green Deal appears to have had a limited effect on carbon reductions. In explaining the failures of the scheme, the then Minister for Energy & Climate Change Ed Davey described it as "too clunky and too complex" and "disappointing" (Vaughan, 2014). The closure of the Green Deal scheme attracted highly critical media headlines: "Green Deal: 'greatest flop of the last Parliament'" (H&V News, 2015), "Disastrous Green Deal is ditched: flagship scheme to insulate homes is branded a £170 million failure" (Cohen, 2015) and "Government kills off flagship Green Deal for home insulation" (Vaughan, 2015). Furthermore, the failure of the Green Deal means that at the time of writing there are no general energy efficiency schemes that households can access (unless they are in a hard-to-treat home or are considered to be vulnerable to fuel poverty), and as a result, improvements to domestic energy efficiency, and by implication carbon savings, are likely to be harder to achieve.

Substantive dimension

One of the major strengths of the Green Deal was its attempt to integrate and stimulate a fragmented energy market. Additionally, it addressed the 'split incentive' problem associated with addressing energy efficiency in the private rented sector (PRS). The PRS is notorious for both having poor energy efficiency and being hard to address through policy given that landlords may be disinclined to make energy efficiency improvements that will provide no direct financial return (ACE, 2011). Peter Mallaburn and Nick Eyre argue that the Green Deal addressed this issue by providing an incentive for private landlords to make energy efficiency improvements, given that there were no upfront costs and the loan was repaid through energy bills (Mallaburn and Eyre, 2014: 38). In social terms as described above, the Green Deal explicitly targeted households that could meet the Golden Rule, largely to ensure that fuel-poor households were not locked into a financial arrangement that led to increased energy costs. In many respects, this was a positive policy approach given that it gave non-fuel-poor households the opportunity to improve the material quality of their homes with associated environmental benefits. However, as Helen Stockton and Ron Campbell (2011) argue, many households are debt adverse, especially those that are already struggling with day-to-day household expenses, and for these groups the Green Deal may not have been an attractive proposition. Equally, for wealthier households, it is argued that they may be more likely to pay for improvements 'up front' rather than go through the complex and binding process of a loan. Additionally, given that the debt was associated with the property not the individual, some argue that this was unappealing to homeowners (Snell and Thomson, 2013). As such, while the Green Deal may have been available to any household meeting the Golden Rule, whether it was sufficiently appealing to different socio-economic groups and different types of households is highly questionable.

Considering the scheme from a substantive dimension, there are four factors likely to have deterred households from participating. First, the design may have acted as a deterrent as the Green Deal was financed through loans that were repaid through energy bills that were bound to the property, rather than to the person who applied for the loan. This meant that households could effectively end up with a loan ascribed to them that they had not taken upon themselves. There was also no guarantee that the costs of the loan would outweigh the costs of repayment, which may have discouraged lower-income households from signing on for a significant financial risk that savings might not pay off (Richards, 2012).

A second deterrent was the upfront assessment cost of up to £150 that households had to pay to be able to proceed with the process (some of which was refunded at a later date). These high fees were seen as a deterrent to lower-income households, instead favouring those with more disposable income (Vaughan, 2013).

A third important deterrent for low-income households was the high interest rate that needed to be repaid: 6.92 per cent, which is higher than most mortgages. This could mean a significant cost, and the failure to repay could mean that households would be disconnected from gas and electricity supplies (The Green Age, 2015).

A fourth important deterrent was the need to have a particular credit score to qualify for the deal. While there were attempts to lower the requirements, this is likely to have further limited take-up (The Green Age, 2015).

Procedural dimension

The Green Deal scheme lacked transparency as hidden costs, the high interest rate and possible penalties meant a financial risk for the households that joined the scheme.

The then Minister for Energy & Climate Change Ed Davey highlighted a lack of information about available measures and "too much red tape", which included preventing customers from being able to take up a plan on the day that they were given a quotation (Pitt, 2014). There was also a lack of clarity within the PRS as to whether responsibility for a Green Deal lay with the landlord or tenant (Pitt, 2014). Furthermore, media reports of 'rogue traders' may have undermined the credibility of the scheme (Green Deal Installer Hub, 2014).

In addition to these failures, the decision to provide public funds to improve scheme take-up has implications in terms of procedural justice. These funds were based on a 'first come, first served' basis, explicitly targeting those not in fuel poverty, paying no attention to means testing and closing once funding had been exhausted, reopening once more funds were available. Consumer websites publicising funding explicitly urged potential applicants to act rapidly; for example, Martin Lewis, a financial journalist, wrote for one website:

> The first £120m in July 2014 went in six weeks, last December another £30m lasted little more than a day ... I think this'll go at super-speed, as many are poised waiting to apply.... It's first come, first served until the cash runs out, so apply asap.
>
> (Lewis, 2015)

Given that to be eligible for the fund households required a Green Deal assessment and quotation from an approved builder, in many respects those successfully accessing the fund were able to do so because they were lucky enough to have the appropriate paperwork as schemes opened. Media reports suggested that even when households had paid for a Green Deal assessment but did not manage to register in time, they lost out on funding (Winch, 2014). This led to criticism over the handling of the scheme by industry (McMullan, 2014) and political opposition, with the shadow minister for Energy & Climate Change, Jonathan Reynolds, writing to the Public Accounts Committee to question whether the speed of applications was driven by "the speculative activity of those seeking to get their hands on as much public subsidy as possible" (Donovan, 2014). The implication here is that many of those applying for scheme vouchers may not have ultimately taken up the free measures, but simply acted to secure available funds.

Given the speed at which schemes opened and then closed, it is questionable whether households with less knowledge about energy efficiency schemes, without internet access or without the capacity to act so quickly would be able to access this support. Furthermore, given the amount of public money ultimately spent on the Green Deal, it is questionable whether the provision of funds in this way was equitable, especially when expenditure is compared to that of the Warm Front Scheme (which specifically targeted energy efficiency improvements for low-income households and had a budget of £210 million between 2011 and 2013 (DECC, 2015b).

Case Study 2: the ECO

The ECO symbolises a shift from financing energy efficiency measures for the poorest through taxes to imposing this role on larger energy suppliers. The ECO requires larger energy suppliers to deliver energy efficiency measures to domestic households. The ECO comprises three elements: the Carbon Emissions Reduction Obligation (CERO), the Carbon Savings Community Obligation (CSCO), and the Home Heating Cost Reduction Obligation (HHCRO) (Ofgem, 2015a). While the CERO addresses hard-to-treat homes, such as those with solid walls, the CSCO and the HCRO focus on energy efficiency improvements in households vulnerable to fuel poverty. The CSCO is area based, focusing on the most deprived communities, and requires 15 per cent of measures to be installed in deprived rural households, whereas the HCRO is based on those in the private housing sector on qualifying benefits (for instance, pension credit or employment-related benefits) (Green Deal Solutions, 2015). The ECO targets are outlined in Table 6.3, and the second round of funding will run until 2017.

The ECO costs approximately £1.3 billion per year (House of Commons, 2015). Approximately £0.54 billion of the £1.3 billion budget is allocated to the two fuel poverty energy efficiency schemes, with a predicted 47 per cent of measures supporting the fuel poor (Snell and Thomson, 2013: 33). This can be compared to substantially higher levels of funding available under the previous government (Hills, 2012).

Table 6.3 ECO targets

Scheme/unit	*1 January 2013–31 March 2015 (ECO)*	*1 April 2015–31 March 2017 (ECO 2)*	*Total*
CERO ($MtCO_2$)	14.0	12.4	26.4
CSCO ($MtCO_2$)	6.8	6.0	12.8
CSCO (rural $MtCO_2$)	1.0	0.9	1.9
HHCRO (£billion)	4.2	3.7	7.9

Source: Ofgem, 2015a.

In July 2014, the government concluded a consultation on the ECO and retrospectively changed several targets and increased the number of allowable energy efficiency measures. Most notably, the CERO dimension of the ECO, targeted towards hard-to-treat homes, was changed, with carbon reduction targets reduced retrospectively by 33 per cent and companies allowed to carry over a proportion of uncompleted work to 2017 (albeit with a penalty rate applied) (Ofgem, 2015b). One of the reasons for this reduction was to reduce the cost borne by energy bill payers by approximately £30–40 per year. In environmental terms, this corresponds with a saving of 14 million tonnes of carbon dioxide compared to the original target of 20.9 million tonnes (Green Installer Hub, 2015). The ECO consultation also led to an increase in the number of rural households eligible for support under the CSCO (from the most deprived 15 per cent of rural Lower Super Output Areas (LSOAs) to 25 per cent), and, perhaps as a result of this, 2015 targets were met; indeed, in support of this point, Ofgem ECO data from November 2014 show negligible progress towards the CSCO rural target, compared to February 2015 data that show the target had almost been reached (Ofgem, 2015c).

In environmental terms, there have been a variety of criticisms levelled at the ECO. In the 2014 ECO consultation, the DECC (2014) reported that 68 per cent of organisations responding to their call disagreed with the proposed 33 per cent reduction for CERO targets. Responding organisations argued that the cut undermined improvements to extremely inefficient housing, lessened the environmental impact of the CERO and generally reduced the funds available to address energy efficiency issues. For example, in their response the ACE argued that "cutting Britain's only national energy efficiency programme – designed to reduce household energy bills and carbon emissions in the long term – to achieve a modest one-off energy bill reduction is completely perverse" (ACE, 2014).

Substantive dimension

One of the challenges of the ECO is the lack of coherence and uniformity in delivering the services to those who need it most. In order to meet their ECO targets, energy companies often work with a variety of stakeholders; for example, British Gas report working with local authorities in Cornwall, Northamptonshire and the North East of England. The Warm Up North scheme, comprising nine local authorities, draws on British Gas ECO funds. Under this scheme, external wall insulation (a measure used for homes that are considered hard to treat) was installed within 40 homes in the Darlington area, with a predicted energy saving of £180 per year (Warm Up North, 2015). Similarly, Npower report working in partnership with Hull City Council to provide external wall insulation to 205 properties as part of a regeneration/anti-fuel poverty development in the Boulevard area of Hull (Npower, 2015).

These two examples highlight the complexity of the delivery of the ECO. On the one hand, individuals who meet qualifying criteria can access energy efficiency improvements through advisory services (run by a range of stakeholders) or directly through energy companies' or the DECC's websites. On the other hand, partnerships such as the two described above mean that some ECO resources are chanelled in a particular way.

This is not to say that such schemes are undesirable, but simply that their results will be highly variable across England, and that eligibility criteria are likely to differ substantially depending upon the partnership in place (a substantial change from the state-funded Warm Front Scheme, which had clear eligibility criteria and was administered at the national level). As some, such as Mallaburn and Eyre (2014), argue, the ECO may compel supplier action, but may not necessarily support the interests of all consumers. As outlined in Table 6.4, the ECO attracts a variety of criticisms in terms of the beneficiaries of its policies (thus raising questions around substantive justice and fuel poverty). The implementation of the ECO has corresponded with a shift to private-sector funding, with virtually no programmes funded through the Treasury. These changes have been widely criticised as regressive (Boardman, 2012; Stockton and Campbell, 2011; Hills, 2012; Snell and Thomson, 2013; CSE, 2014) given that the ECO is funded through domestic energy bills and was predicted to increase these by £40–60 per year (Climate Change Committee, 2011).

The CERO is especially controversial given its focus on hard-to-treat properties. Indeed, the ACE describes the funding mechanism as a regressive "double dividend whereby [potentially wealthy] households may be able to access expensive energy efficiency improvements and reduced bills, whereas other [potentially fuel poor] households may not be eligible for support, and have increased bills as a result" (ACE, 2014). Implicit within this is that some households may be pushed (further) into fuel poverty and exposed to a variety of negative social and health outcomes as a result (see Table 6.1).

The CSCO and the HHCRO are also subject to similar criticisms. The CSE argues that the HHCRO is likely to worsen fuel poverty given that it funds high-cost energy efficiency improvements to a small number of households (approximately 500,000) while also raising energy bills (CSE, 2014; ACE, 2014). The CSE argues that the HHCRO is regressive, unlike its Treasury-funded predecessor. The CSE goes on to argue that "it is difficult to believe the HHCRO was introduced for any other reasons than to remove the cost from public spending and hope to 'hide' it on fuel bills" (CSE, 2014: 28). Furthermore, the ACE argues that the "HHCRO ... has in England failed to deliver outcomes that match the new fuel poverty strategy's targets of reaching certain EPC bands by deadline dates" (ACE, 2014: 2). By only generally delivering one measure per home and not supporting households to access further energy efficiency improvements, the scheme fails to improve the energy efficiency of households sufficiently

Table 6.4 Analytical framework

Policy	*Substantive dimensions*	*Procedural dimension*	*Social and environmental policy dimension*
Green Deal	The Green Deal offered the opportunity for householders to improve the physical build of their property without an upfront cost. This had the potential to improve living conditions and thermal comfort. However, as a programme of work explicitly aimed at the non-fuel poor, benefits would have been felt by those in the easiest-to-treat properties who were less likely to be suffering from the effects of low temperatures.	The Green Deal could only be taken up by those likely to meet the 'Golden Rule'. This was designed to support those in hard-to-treat homes and the fuel poor. In order to stimulate take-up, publicly funded financial incentives were provided. These were made on a 'first come, first served' basis rather than according to means testing/qualifying criteria.	The Green Deal was regarded as innovative in its ambition as it tried to unite a highly fragmented market. However, in reality, the take-up of the Green Deals was very low, and as such, its environmental impact was limited. The use of public money to incentivise take-up seems unfair when compared to the substantial reductions in the budgets of schemes targeting energy efficiency measures in fuel-poor homes.
CERO element of ECO	CERO costs are added to all energy bills. This is criticised by many as regressive, with costs being passed on to both wealthy and poor households alike.	Access to the CERO is not means tested but is instead based on the physical build of the property. Access to the ECO varies nationally. While some schemes may be accessed directly, there is substantial variation in schemes across local areas and regions. Awareness of the ECO is low; this is exacerbated by complex eligibility criteria and the nature of fuel-poor households (for instance, levels of fuel poverty among poor older people are high, and it is this group that are most unlikely to access information through the internet, where most schemes are advertised).	Energy companies are typically driven by the need to meet targets, and are likely to do so in the manner that is the most efficient rather than the most socially just. This issue can be addressed by the partnerships created through the ECO. Typically, local authorities have a good understanding of local issues (such as housing problems and poverty) and are well positioned to target those most in need. However, these partnerships vary greatly and rely upon local action and expertise.
CSCO and HHCRO elements of ECO	While the CSCO and the HHCRO target those vulnerable to fuel poverty, the decision to fund these through energy bills is still questioned by many. Over and above this, overall budgets available for the fuel poor have been reduced substantially, implying that less support for those most in need is available.		

(ACE, 2014). The CSCO, the deprived rural aspect of the ECO, has also proven difficult to fulfil, with the CSE finding that in the first year only 7.5 per cent of households receiving measures were in rural areas (CSE, 2014).

In environmental terms, the ACE argues that existing policy has led to "an ever-widening gap between the carbon savings required from the residential sector under the Climate Change Act and the carbon savings that will be delivered" (ACE, 2014). In addition to criticisms regarding the reduced targets, others argue that the flexibility of the ECO requirements means that suppliers inevitably opt for the cheapest options in order to meet their requirements, which are not always those with the greatest carbon savings (CSE, 2014) or with the most desirable social outcomes (Mallaburn and Eyre, 2015). Unlike the Green Deal, a substantial programme of work has occurred through the ECO. However, whether the scheme is sufficiently ambitious in terms of either the number of measures delivered or the overall carbon savings is questioned by some (for example, ACE, 2014), especially in terms of its capacity to deal with 'deep retrofits' and harder improvements such as solid wall insulation.

Procedural dimension

There are also questions around procedural justice, most notably about who has been included within ECO schemes. DECC policy suggests that the ECO enables 'well-placed' energy companies to engage with their customers (and beyond) and thus deliver measures in an effective way. However, many argue that the sector is inexperienced at delivering such widespread energy efficiency measures to those vulnerable to fuel poverty, especially when compared to local authorities and landlords (CSE, 2014). One consequence of this has been that energy companies have been accused of targeting easy wins, the so-called 'low-hanging fruit' (Politics Home, 2015), rather than those in most need (both socially and environmentally). One way of addressing this gap in knowledge has been to develop a variety of partnerships and networks in order to deliver ECO projects. While a range of innovative projects have been delivered, the implication is that a proportion of ECO funding has been used in an ad hoc manner led by the priorities of social landlords, local authorities or other stakeholders (although constrained by ECO guidelines on acceptable measures). The result of this is likely to be highly patchwork in nature (unlike the Warm Front Scheme), and while the ECO itself was subject to a national consultation, many ECO schemes delivered through partnerships are unlikely to be subject to any democratic decision-making processes.

A further point surrounds enrolment procedures. While some schemes have proactively targeted hard-to-treat houses in areas of socio-economic deprivation (for example, the Yorkshire Energy Partnership has sought to treat entire streets of houses without cavity walls) and households have been approached with the offer of free measures, this varies greatly and is entirely dependent upon the partnerships and schemes in place. Indeed, where local partnerships do not exist, eligible households may be expected to access ECO information from the websites of energy companies or the DECC (as opposed to being actively informed or approached about measures). As such, even if a household is eligible for a measure, whether they are aware of this, are able to access sources of information or have the capacity to apply for it is questionable. Given the relationship between fuel poverty, old age, poverty, disability and other indicators of vulnerability, whether households have access to the internet in order to research schemes or a general awareness of available support must be questioned, and where schemes are publicised in a passive manner (for instance, on a government website), the most vulnerable households may struggle to seek this information out. For example, research conducted with food bank users living in ECO-eligible households in 2015 found very limited knowledge about support on offer, and no take-up of available measures (Lambie-Mumford and Snell, 2015).

Social implications, challenges and solutions

Despite its promising start, ultimately the way in which the Green Deal was developed made it unpalatable to consumers, and resulted in very limited energy efficiency measures and associated carbon savings. At the national level, while the Green Deal was not targeted towards the fuel poor, the public funds that were ultimately spent in support of the programme were targeted in an indiscriminate manner. Given the decimation of public funds previously used in support of improving homes of those vulnerable to fuel poverty, serious questions of both distributive and procedural justice are raised.

As described above, while the ECO has been more successful in delivering carbon reductions, whether the housing stock is being sufficiently transformed is questionable. ECO carbon targets have been lowered, in part to reduce the burden on energy bill payers. While many have criticised the funding of the ECO as being regressive, most would argue that it is not the level of investment that is the problem but actually the source of the investment. As Mallaburn and Eyre argue:

> some ministers, regulators and public officials have periodically claimed that market forces on their own will deliver all cost-effective energy efficiency savings. This is simply not true ... these claims tend to be driven by ideological assumptions rather than by any serious examination of the scientific evidence.
>
> (Mallaburn and Eyre, 2014: 35)

Indeed, organisations such as the ACE and the CSE argue that the fuel poverty dimensions of energy efficiency should be funded separately by the state to ensure the social sustainability of such schemes (ACE, 2014).

At the time of writing, the government is conducting a consultation on the future of energy efficiency policy. While the Green Deal has already ended, the ECO will close in 2017. The Energy & Climate Change Committee have set three questions for organisations responding to the consultation:

1 Why have previous approaches to energy efficiency failed to deliver significant results?
2 What lessons can be learnt from current and previous schemes including the Green Deal, the Green Deal Home Improvement Fund and the ECO?
3 How does the UK's performance on home energy efficiency compare with other countries? What lessons can be learned from these countries on energy efficiency? (Disability Rights UK, 2015).

The phrasing of these questions implies recognition among policy-makers that existing schemes have been limited in impact.

Table 6.5 summarises the key challenges associated with domestic energy policy. While domestic energy efficiency schemes have the potential to simultaneously address social and environmental needs, if they are not designed appropriately they may actually worsen fuel poverty, for example, through regressive funding mechanisms that place a disproportionate financial burden on the poorest households. Additionally, where policies are designed to encourage households to improve the energy efficiency of their homes at their own risk (such as the Green Deal), there has to be sufficient incentive to do so. As implied by the consultation questions described above, understanding where existing UK energy efficiency policy has failed will be crucial for future policy development. However, existing criticisms strongly point towards addressing the regressivity of programmes, considering the barriers that prevented households from taking up a Green Deal and ensuring that the most vulnerable customers are both aware of and able to access schemes.

In conclusion, balancing the environmental and social aims implicit in most energy policies is a highly complex task. While addressing the global commitment to reduce carbon emissions, the most vulnerable in England should also be protected from the effects of fuel poverty. Coalition policy generally, and the Green Deal and the ECO specifically, appear neither ambitious in terms of carbon reductions nor sufficiently focused on the needs of the most vulnerable.

Table 6.5 Social implications, challenges and solutions

Social implications	*Challenges*	*Solutions*
Balancing the social and environmental objectives implicit within domestic energy policy is a highly challenging task. If carbon savings are not made within the domestic energy sector, there are a variety of social implications at the global level associated with climate change. However, at the national level, it is essential that the living conditions of those most vulnerable to fuel poverty are addressed. Without this, householders face an increased risk of poor health, social exclusion and other forms of socio-economic deprivation.	Schemes such as the Green Deal rely upon consumer willingness to improve the energy efficiency of their home with no immediate financial gain. For ECO-related schemes, funding is a key challenge. Where energy efficiency improvements are funded through energy bills, these may actually worsen fuel poverty. Access to schemes is also a key concern, especially given the variety of schemes and the likely variation in eligibility criteria (especially for area-based programmes). Whether the most vulnerable households are aware of schemes, or are able to access information, is also questionable.	A voluntary programme such as the Green Deal needs to be sufficiently attractive to customers, and as many have argued, reducing the interest rate and encouraging more engagement from trusted brand names may encourage take- up (Snell and Thomson, 2013). A key criticism of existing policy is the lack of state funding. Schemes funded through the Treasury are typically regarded as more progressive, given that energy efficiency improvements can be made without increasing energy bills. This ensures that the poorest households do not spend a disproportionate amount of their income on increased energy bills (when compared to the wealthiest households). While many local ECO-funded schemes have proved innovative in terms of both energy efficiency and anti-fuel poverty measures, more proactive promotion of eligibility for schemes at the national level may be desirable. Lessons should be drawn from the Warm Front Scheme, which was advertised nationally, had clear branding and a specific website and was signposted by a variety of government departments and third-sector organisations.

Notes

1 It should be noted here that two key policy levels are discussed within this chapter: policy relating to the UK as a whole, and specific policies implemented within England.

2 It should be noted here that fuel poverty is a partially devolved policy issue, with the four nations having individual policy targets and measurements (DECC, 2014). While Wales and Northern Ireland have very similar measures in place, the policy detail outlined here relates to England.

3 The same shift in responsibility has not occurred in the devolved regions.

References

ACE (Association for the Conservation of Energy) (2011), "Written evidence submitted by the Association for the Conservation of Energy to the Climate Change Committee", House of Commons Energy & Climate Change Committee, available at: www.publications.parliament.uk/pa/cm201012/cmselect/cmenergy/1744i_ii/1744we12.htm (accessed 26 March 2016).

ACE (Association for the Conservation of Energy) (2012), *Dead CERT: Framing a Sustainable Transition to the Green Deal and the Energy Company Obligation*, London: ACE, available at: www.ukace.org/wp-content/uploads/2012/01/ACE-Research-2012-01-Dead-CERT-full-report.pdf (accessed 26 March 2016).

ACE (Association for the Conservation of Energy) (2013), *Fact-File: The Cold Man of Europe*, London: ACE, available at: www.ukace.org/wp-content/uploads/2013/03/ACE-and-EBR-fact-file-2013-03-Cold-man-of-Europe.pdf (accessed 26 March 2016).

ACE (Association for the Conservation of Energy) (2014), *The Future of the Energy Company Obligation*, London: ACE, available at: www.ukace.org/wp-content/uploads/2014/04/ACE-Consultation-Response-2014-04-The-Future-of-the-Energy-Company-Obligation.pdf (accessed 26 March 2016).

BBC (2015), "Green Deal funding to end, government announces", available at: www.bbc.co.uk/news/uk-england-gloucestershire-33638903 (accessed 24 March 2016).

Boardman, B. (1991), *Fuel Poverty: From Cold Homes to Affordable Warmth*, London: Belhaven Press.

Boardman, B. (2012), "Fuel poverty synthesis: lessons learnt, actions needed", *Energy Policy*, 49, 143–8.

Cahill, M. (2001), *The Environment and Social Policy*, London: Routledge.

Chester, L. (2010), "Conceptualising energy security and making explicit its polysemic nature, *Energy Policy*, 38(2), 887–95.

Climate Change Committee (2011), "Household energy bills: impacts of meeting carbon budgets", available at: www.theccc.org.uk/publication/household-energy-bills (accessed 22 June 2016).

Climate Change Committee (2015), "The Climate Change Act and UK regulations", available at: www.theccc.org.uk/tackling-climate-change/the-legal-landscape/global-action-on-climate-change/ (accessed 26 March 2016).

Consumer Focus (2011), *Access for All*, London: Consumer Focus.

Cohen, T. (2015), "Disastrous Green Deal is ditched: flagship scheme to insulate homes is branded a £170 million failure", *Daily Mail*, 24 July, available at: www.dailymail.co.uk/news/article-3172844/Disastrous-Green-Deal-ditched-Flagship-scheme-insulate-homes-branded-170million-failure.html#ixzz3jiuo4n15 (accessed 24 March 2016).

CSE (Centre for Sustainable Energy) (2014), *Beyond the ECO: An Exploration of Options for the Future of a Domestic Energy Supplier Obligation*, Bristol: CSE, available at: www.cse.org.uk/downloads/file/beyond-the-ECO.pdf (accessed 26 March 2016).

Davey, E. (2014), "Speech by Energy Secretary, Edward Davey MP, to the *Economist* Energy Summit on the UK's energy security", available at: www.gov.uk/government/speeches/uk-energy-security-active-government-smart-intervention (accessed 26 March 2016).

DECC (Department of Energy & Climate Change) (2010), *The Green Deal: A Summary of the Government's Proposals*, London: DECC, available at: www.gov.uk/government/uploads/system/uploads/attachment_data/file/47978/1010-green-deal-summary-proposals.pdf.

DECC (Department of Energy & Climate Change) (2013), *United Kingdom Housing Energy Fact File, 2013*, London: DECC, available at www.gov.uk/government/uploads/system/uploads/attachment_data/file/345141/uk_housing_fact_file_2013.pdf (accessed 26 March 2016).

DECC (Department of Energy & Climate Change) (2014), *The Future of the Energy Company Obligation: Government Response to the 5 March 2014 Consultation*, London: DECC, available at: www.gov.uk/government/uploads/system/uploads/attachment_data/file/342178/The_Future_of_the_Energy_Company_Obligation_Government_Response.pdf (accessed 26 March 2016).

DECC (Department of Energy & Climate Change) (2015a), *Cutting the Cost of Keeping Warm: A Fuel Poverty Strategy for England*, London: DECC, available at www.gov.uk/government/publications/cutting-the-cost-of-keeping-warm (accessed 26 March 2016).

DECC (Department of Energy & Climate Change) (2015b), *2010 to 2015 Government Policy: Household Energy*, London: DECC, available at www.gov.uk/government/publications/2010-to-2015-government-policy-household-energy/2010-to-2015-government-policy-household-energy#bills-and-legislation (accessed 26 March 2016).

DECC (Department of Energy & Climate Change) (2015c), *Evaluation of the Big Energy Saving Network: Final Report*, London: DECC, available at: www.gov.uk/government/uploads/system/uploads/attachment_data/file/405800/BESN_final_reporl.pdf (accessed 26 March 2016).

DECC (Department of Energy & Climate Change) (2015d), *The Central Heating Fund: Guidance for Local Authorities*, London: DECC, available at: www.gov.uk/government/uploads/system/uploads/attachment_data/file/417631/Central_Heating_Fund_LA_guidance.pdf (accessed 26 March 2016).

DECC (Department of Energy & Climate Change) (2015e), "The heat is on to ensure warmer, healthy homes for everyone", available at: www.gov.uk/government/news/the-heat-is-on-to-ensure-warmer-healthy-homes-for-everyone (accessed 26 March 2016).

DEFRA (Department for Environment, Food & Rural Affairs) (2012), *UK Climate Change Risk Assessment: Government Report*, London: DEFRA, available at www.gov.uk/government/publications/uk-climate-change-risk-assessment-government-report (accessed 26 March 2016.

Disability Rights UK (2015), "New inquiry into home energy efficiency", available at: www.disabilityrightsuk.org/news/2015/september/new-inquiry-home-energy-efficiency#sthash.iLnJsIke.dpuf (accessed 24 March 2016).

Donovan, N. (2014), "Pay delay hits Green Deal providers", *BBC News*, 11 August, available at: www.bbc.co.uk/news/business-28744792 (accessed 24 March 2016).

Dowson, M., Poole., A., Harrison, D. and Susman, G. (2012), "Domestic UK retrofit challenge: barriers, incentives and current performance leading into the Green Deal", *Energy Policy*, 50, 294–305.

Duxbury, N. (2012), "Green Deal changes to benefit low-income homes", *Inside Housing*, 11 June, available at: www.insidehousing.co.uk/green-deal-changes-to-benefit-low-income-homes/6522284.article (accessed 24 March 2016).

Ellis, M. (2014), "Green Deal Home Improvement Fund launches", *USwitch*, 10 June, available at: www.uswitch.com/gas-electricity/news/2014/06/10/green-deal-home-improvement-fund-launches/ (accessed 26 March 2016).

Energy Saving Trust (2015), "Smart meters & controls", available at: www.energysavingtrust.org.uk/domestic/smart-meters-controls (accessed 26 March 2016).

Fitzpatrick, T. (ed.) (2012), *Understanding the Environment and Social Policy*, Bristol: Policy Press.

The Green Age (2015), "Why did the green deal fail?", available at: www.thegreenage.co.uk/why-did-the-green-deal-fail/ (accessed 26 March 2016).

Green Deal Initiative (2015), "What is the Green Deal?", available at: www.greendealinitiative.co.uk/about-the-green-deal/the-green-deals-golden-rule (accessed 26 March 2016).

Green Deal Installer Hub (2014), "Green Deal rouge [sic] traders", 15 January, available at: www.greendealinstallerhub.co.uk/148,news,green_deal_rouge_traders.html (accessed 24 March 2016).

Green Deal Installer Hub (2015), "New ECO consultation released", 11 March, available at: www.greendealinstallerhub.co.uk/158,news,new_eco_consultation_released.html (accessed 26 March 2016).

Green Deal Solutions (2015), "ECO funding: Energy Companies Obligation", available at: www.greendealsolutions.net/eco-funding (accessed 26 March 2016).

Guertler, P. (2012), "Can the Green Deal be fair too? Exploring new possibilities for alleviating fuel poverty", *Energy Policy*, 49, 91–7.

Gupta, R., Barnfield, L. and Hipwood, T. (2014), "Impacts of community-led energy retrofitting of owner-occupied dwellings", *Building Research and Information*, 42(4), 446–61.

Hamilton, I., Shipworth, D., Summerfield, A., Steadman, P., Oreszczyn, T. and Lowe, R. (2014), "Uptake of energy efficiency interventions in English dwellings", *Building Research and Information*, 42(3), 255–75.

Hills, J. (2012), *Getting the Measure of Fuel Poverty: Final Report of the Fuel Poverty Review*, London: Centre for Analysis of Social Exclusion, available at: http://sticerd.lse.ac.uk/dps/case/cr/CASEreport72.pdf (accessed 26 March 2016).

House of Commons (2014), *Smart Meters*, House of Commons Briefing Note, London: HMSO, available at: researchbriefings.files.parliament.uk/documents/SN06179/SN06179.pdf (accessed 26 March 2016).

House of Commons (2015), *ECO, the Energy Company Obligation*, House of Commons Briefing Note, London: HMSO, available at: researchbriefings.files.parliament.uk/documents/SN06814/SN06814.pdf (accessed 26 March 2016).

House of Commons Committee of Public Accounts (2011), *Preparations for the Roll-Out of Smart Meters*, London: HMSO, available at: www.publications.parliament.uk/pa/cm201012/cmselect/cmpubacc/1617/1617.pdf (accessed 26 March 2016).

Huby, M. (1998), *Social Policy and the Environment*, Buckingham: Open University Press.

H&V News (2015) "Green Deal: 'greatest flop of the last Parliament'", 27 July, available at: www.hvnplus.co.uk/news/green-deal-greatest-flop-of-the-last-parliament/8686791.article (accessed 24 March 2016).

Kerley, R. (2012), "Green economy, carbon and energy editorial", *Scottish Policy Now*, December, available at: www.scottishpolicynow.co.uk/article/greeneditorial (accessed 26 March 2016).

Lambie-Mumford, H. and Snell, C. (2015), *Heat or Eat: Food and Austerity in Rural England*, Leeds: Communities & Culture Network, available at www.communitiesandculture.org/files/2015/03/Lambie-Mumford-and-Snell-_-Food-and-Austerity-in-Rural-England-Interim-Report-March-2015.pdf (accessed 26 March 2016).

Lewis, M. (2015), "Urgent: get up to £5,600 to improve your home", *Moneysavingexpert.com*, 11 March, available at: www.moneysavingexpert.com/news/utilities/2015/03/urgent-get-up-to-5600-to-improve-your-home (accessed 26 March 2016).

McCauley, D., Heffron, J. R., Stephan, H. and Jenkins, K. (2013), "Advancing energy justice: the triumvirate of tenets", *International Energy Law Review*, 32(3), 1–5.

McMullan, T. (2014), "Green Deal hit by concerns over unclaimed cashback vouchers", *GreenWise*, 5 August, available at: http://greenwisebusiness.co.uk/news/green-deal-hit-by-concerns-over-unclaimed-cashback-vouchers-4425.aspx#.ViZVR7Q-BBw (accessed 24 March 2016).

Mallaburn, P. and Eyre, N. (2014), "Lessons from energy efficiency policy and programmes in the UK from 1973 to 2013", *Energy Efficiency*, 7(1), 23–41.

Marmot Review Team (2011), *The Health Impacts of Cold Homes and Fuel Poverty*, London: Friends of the Earth/Marmot Review Team, available at: www.foe.co.uk/sites/default/files/downloads/cold_homes_health.pdf (accessed 26 March 2016).

NICE (National Institute for Health and Care Excellence) (2015), "Excess winter deaths and illness and the health risks associated with cold homes", available at: www.nice.org.uk/guidance/ng6 (accessed 26 March 2016).

Npower (2015), "We help 205 properties in Hull become more energy efficient as part of the award-winning Boulevard Improvement Scheme", available at: www.npower.com/home/energy-efficiency/home-energy-saving-schemes/our-partners/hullcitycouncil/ (accessed 26 March 2016).

Ofgem (2015a), "Warm Home Discount (WHD)", available at: www.ofgem.gov.uk/environmental-programmes/social-programmes/warm-home-discount (accessed 26 March 2016).

Ofgem (2015b), *Energy Companies Obligation (ECO) Compliance Update: June 2015*, London: Ofgem, accessed at: www.ofgem.gov.uk/ofgem-publications/95225/energycompaniesobligationecocomplianceupdate-june2015-pdf (accessed 26 March 2016).

Ofgem (2015c), *Energy Companies Obligation (ECO) Compliance Update: February 2015*, London: Ofgem, available at: www.ofgem.gov.uk/sites/default/files/docs/2015/02/eco_compliance_update_february_2015_v2_0.pdf (accessed 26 March 2016).

Pitt, V. (2014), "Davey promises to fix 'clunky' Green Deal", *BDonline*, 6 March, available at: www.bdonline.co.uk/news/davey-promises-to-fix-clunky-green-deal/5067044.article (accessed 24 March 2016).

Politics Home (2015), "Energy Company Obligation", available at: www.politicshome.com/organisation/page/calor-gas/energy-company-obligation (accessed 26 March 2016).

Pollitt, M. G. (2012), "The role of policy in energy transitions: lessons from the energy liberalization era", Electricity Policy Research Group Working Paper 1208, University of Cambridge.

Richards, J. (2012), "The Green Deal", *Mortgage Finance Gazette*, 6 December, available at: www.mortgagefinancegazette.com/legal/the-green-deal/ (accessed 26 March 2016).

Select Committee on Energy & Climate Change (2014), "The Green Deal: watching brief", available at: www.publications.parliament.uk/pa/cm201415/cmselect/cmenergy/348/34805.htm (accessed 26 March 2016).

Snell, C. and Haq, G. (2014), *The Short Guide to Environmental Policy*, Bristol: Policy Press.

Snell, C. and Thomson, H. (2013), "Energy policy under the Coalition government: Green Deal or no deal", *Social Policy Review*, 25, 23–46.

Snell, C., Bevan, M. and Thomson, H. (2015), "Justice, fuel poverty and disabled people in England", *Energy Research and Social Science*, 10, 123–32.

Sovacool, B. K. and Dworkin, H. M. (2014), *Global Energy Justice: Problems, Principles, and Practices*, Cambridge: Cambridge University Press.

Sovacool, B. K. and Mukherjee, I. (2011), "Conceptualising and measuring energy security: a synthesized approach", *Energy Policy*, 5343–55.

Sovacool, B. K., Sidortsov, R. V. and Jones, B. R. (2014), *Energy Security, Equality, and Justice*, London: Routledge.

Stockton, H. and Campbell, R. (2011), *Time to Reconsider UK Energy and Fuel Poverty Policies?*, York: JRF, available at: www.jrf.org.uk/report/time-reconsider-uk-energy-and-fuel-poverty-policies (accessed 26 March 2016).

UNFCCC (United Nations Framework Convention on Climate Change) (2012), "First steps to a safer future: introducing the United Nations Framework Convention on Climate Change", available at: http://unfccc.int/essential_background/convention/items/6036.php (accessed 26 March 2016).

UNFCCC (United Nations Framework Convention on Climate Change) (2013), "Feeling the heat: climate science and the basis of the Convention", available at: http://unfccc.int/essential_background/the_science/items/6064.php (accessed 26 March 2016).

Vaughan, A. (2013), "Green Deal's upfront fees 'put people off upgrading homes'", *Guardian*, 6 January, available at: www.theguardian.com/environment/2013/jan/06/green-deal-upfront-fees-upgrading (accessed 26 March 2015).

Vaughan, A. (2014), "Green Deal loan take-up is 'disappointing', Ed Davey concedes", *Guardian*, 5 March, available at: www.theguardian.com/environment/2014/mar/05/green-deal-loan-take-up-disappointing-ed-davey-eco (accessed 24 March 2016).

Vaughan, A. (2015), "Government kills off flagship Green Deal for home insulation", *Guardian*, 23 July, available at: www.theguardian.com/environment/2015/jul/23/uk-ceases-financing-of-green-deal (accessed 24 March 2016).

Walker, G. and Day. R. (2012), "Fuel poverty as injustice: integrating distribution, recognition and procedure in the struggle for affordable warmth", *Energy Policy*, 49, 69–75.

Warm Up North (2015), "Warm Up North", available at: warmupnorth.com (accessed 26 March 2016).

Wilkinson, P., Smith, K., Joffe, M. and Haines, A. (2007), "A global perspective on energy: health effects and injustices", *The Lancet*, 370(9591), 15–21.

Winch, J. (2014), "Was Green Deal cashback scheme rigged?", *Daily Telegraph*, 2 August, available at: www.telegraph.co.uk/finance/personalfinance/energy-bills/11006180/Was-Green-Deal-cashback-scheme-rigged.html (accessed 24 March 2016).

7 Pro-poor technology transfer

The case of solar power and refugees in Africa

Johan Nordensvärd with Frauke Urban

The social dimensions of access to low-carbon energy

One important challenge of low-carbon development is to increase access to low-carbon energy technology for the world's poorest people. Most low-carbon innovation has been developed, designed, built and deployed in high-income countries and increasingly also in middle-income countries such as China, India and Brazil. The world's poorest countries have only made a minor contribution to greenhouse gas emissions leading to climate change. Least Developed Countries (LDCs) are reported to have accounted for only about 4 per cent of global greenhouse gas emissions in 2005 and only 0.3 per cent of historic carbon emissions from energy use (WRI, 2005), yet they are the most vulnerable to the impacts of climate change (IPCC, 2007, 2014). It is therefore less a question of mitigating emissions leading to climate change and much more a question of exploring the opportunities and benefits that low-carbon development can have for the poorest and most vulnerable people. While there may be a risk that climate change mitigation could harm development in poor countries (Jakob and Steckel, 2013) there are also opportunities for linking low-carbon development with poverty reduction strategies, including through the transfer of low-carbon energy technologies. Hence, technology transfer needs to be pro-poor.

Poverty reduction is seen as

> a complex process that requires several strategies such as improved access to funding, human capital and technologies, as well as encouraging good governance, equal distribution of resources and services – such as education, employment, health care and energy access – and encouraging development.
>
> (Urban *et al.*, 2013: 230)

Often poverty reduction and low-carbon development have been seen as competing interests as fuel poverty has often been fought in developing countries by increasing public spending on fossil fuel subsidies. There is an argument that fossil fuel subsidies are taking up more and more of states' budgets. "[A]ttempts to remove subsidies typically meet strong opposition, often from poorer urban populations" and "poor people are proportionately more affected by subsidy removal" (Lockwood, 2013: 27). In developing countries, people are doubtful that reducing fossil fuel subsidies will translate into more funding for pro-poor welfare. At the same time, access to electricity is seen as a prerequisite for development (Forsyth, 1999; Juma and Yee-Cheong, 2005; Modi *et al*., 2005) so there is a particular need to decrease the dependence upon fossil fuels by transitioning to sustainable low-carbon energy, particularly renewable energy.

This has become even more important in places where people, particularly women and girls, have to collect fuelwood for cooking and/or for heating their homes. These chores tend to take much of their time and, as shown in this case study, they can also pose a danger. Wahidul Biswas and colleagues (2001) argue that low-carbon energy could free up time for these people and improve their socio-economic situation. This is an urgent problem as 1.3 billion people lack access to electricity and 2.7 billion people use traditional biomass, such as fuelwood, dung and agricultural residue for cooking and heating (IEA, 2011). Improvements in reducing energy poverty have been slow for decades and several years ago, the International Energy Agency (IEA) predicted that even in 2030 more than 1 billion people will not have access to electricity. Nevertheless, these figures have now been revised due to the introduction of the Sustainable Energy for All initiative and the energy targets of the Sustainable Development Goals, which aim for universal energy access for everyone by 2030. It remains to be seen whether these targets will be achieved in the coming two decades. There is a global consensus that access to modern energy, including electricity and clean cooking fuels, is important for improving people's standard of health, education, food security, gender equality and ability to earn a living (DFID, 2002). One of the main advantages of renewable energy technology based on solar, wind, biomass and small hydro, is its distributed nature, which can make it appealing to off-grid households and communities. Displaced people such as refugees and those living in areas where the state fails to deliver energy services, such as areas under the control of armed groups, can benefit from the decentralised nature of renewable energy technology (Urban and Lind, 2011). Specific solar energy technologies like solar cookers, solar lamps and photovoltaic (PV) panels have proven useful for off-grid communities as well as for displaced communities in recent years.

One of the largest challenges has been to address how low-carbon technology transfers could reach some of the poorest people in the least developed countries in the world. Most Western and Eastern countries and corporations have developed and deployed large-scale technologies such as hydropower dams and large-scale wind and solar energy farms, which contrast with the needs of the poorest communities in LDCs who lack capital and spending power to make investments in low-carbon energy technology profitable in the long run.

There tends to be a conflict between elitist decision-makers' wish for "rapid, large-scale deployment" (Ockwell and Mallett, 2013: 123) and the imperative to find technologies that fit the needs of poorer communities, such as refugees displaced by conflict or natural disasters. Some scholars argue for a new approach that will respond to the overall situation of poor communities and there is an need "to understand and respond to the needs, desires and existing cultural practices of local stakeholders whose engagement is critical to the uptake of new low carbon technologies" (Ockwell and Mallett, 2013: 123). Contentious issues are not only scale, but also which actors should take responsibility for technology transfer to poorer vulnerable communities in LDCs. Transferring low-carbon technologies is more than just a focus on hardware and finance (Byrne *et al.*, 2011); there also needs to be the capacity to absorb new technologies in local communities (Cohen and Levinthal, 1990). It is important that local communities have certain levels of knowledge and skills to manage and operate the transferred technologies, as well as shaping links to national and local institutions to benefit from low-carbon technologies (Bell and Pavitt, 1993; Bell, 2009).

Technology transfer and low-carbon energy

Carbon markets and technological transfer have become dominant concepts in the global discussion on climate change. The UN has focused much on the importance of making low-carbon technology available to developing countries. Both the UNFCCC and the Kyoto Protocol enforce the obligation of developed nations to support and finance the transfer of low-carbon technology to low- and middle-income countries (UNFCCC, 1992). Technology transfer played an important role in engaging developing countries to participate in international policy platforms such as the UNFCCC and it is reiterated that access to important low-carbon technology would benefit these countries both economically and socially (Ockwell and Mallett, 2013). Low-carbon technological transfer is therefore an attractive concept for donor organisations, intergovernmental organisations and NGOs.

Low-carbon technology transfer is supposed to create win-win situations. For example, through the Clean Development Mechanism (CDM), Annex 1 countries can offset some of their emissions through sharing or financing climate-relevant technology in non-Annex 1 countries. For emerging and developing economies, this could mean that technology can be transferred by a firm, government, NGO or academic institution. Technology could also be owned or produced by a domestic actor (Forsyth, 1999). The nature of technological transfer is in reality diverse and could mean short- or long-term transfer, transfers within a larger corporation, joint ventures or just the selling of the technology. It could range from formal agreements to more informal means such as personnel movement, networks and publication (Ockwell and Mallett, 2013).

Table 7.1 Environmental and social implications of technology transfer

Environmental implications	*Social implications*	*Social sustainability*	*Social policy*
Low-carbon technology transfer can help to achieve a transition from fossil fuels to renewable energy, and can thereby contribute to climate change mitigation. At the same time, the rise of certain types of energy technology, such as large hydropower dams, through the Clean Development Mechanism (CDM) and other technology transfer deals can lead to adverse environmental and social impacts. Reducing traditional biomass such as fuelwood and replacing it with renewable energy such as solar cookers has potentially large environmental benefits, such as reducing the pressure on wood and forests, thereby decreasing deforestation, protecting carbon sinks and safeguarding biodiversity.	Providing access to clean, affordable modern energy such as from solar PV, solar lamps or solar cookers has the potential to make a huge positive impact on the lives of people living in fuel poverty. It has positive impacts on health due to reduced indoor air pollution, providing lighting for studying, increasing safety for women and girls by reducing the need to collect fuelwood and offering potential opportunities for income generation and enterprise activities.	The social sustainability of technology transfer depends to some extent upon the type of technology and its implementation. For example, large hydropower dams may lead to severe unsustainable consequences such as displacement and resettlement, threats to livelihoods, inadequate compensation, and so forth. Other types of low-carbon technology transfer can be very beneficial and socially sustainable for local people, such as small-scale solar energy or wind energy technology. Long-term social sustainability depends upon a complex set of issues including the scale, whether the technologies are fit for purpose for the recipients, the support offered by the donors/firms/governments providing the technologies, the ability to maintain them, whether or not they are locally sourced and produced, and so forth.	There are many opportunities for social policy to use low-carbon energy technology as a means to promote sustainable development and poverty reduction. For example, this could be through providing access to modern energy, improving people's health by reducing indoor air pollution, encouraging women's and girls' empowerment, creating income generation opportunities, and so forth. The challenge is how to achieve this when states are weak or even failing, and when donor and private-sector activity is patchy and ad hoc.

The IPCC's Special Report *Methodological and Technological Issues in Technology Transfer* defines 'technology transfer' as a

> broad set of processes covering the flows of know-how, experience and equipment for mitigating and adapting to climate change [...]. The broad and inclusive term 'transfer' encompasses diffusion of technologies and technology cooperation across and within countries. It comprises the process of learning to understand, utilise and replicate the technology, including the capacity to choose it and adapt it to local conditions and integrate it with indigenous technologies.
>
> (Hedger *et al.*, 2000: 1.4)

This definition thus includes both technology transfer and technology cooperation. The understanding of technology transfer and cooperation has broadened in recent years, moving away from pure "hardware-thinking" (Urban *et al.*, 2015: 235–6). Technology transfer involves flows of know-how, information, experience and equipment, either from North to South, South to North, North to North or South to South.

> It can be distinguished between three flows within the process of technology transfer and technology cooperation: (1) capital goods and equipment (2) skills and know-how for operation and maintenance (3) knowledge and expertise for innovation (Ockwell *et al.*, 2007, 2010; Ockwell and Mallett, 2012, 2013). Technology transfer and technology cooperation can be between firms, hence horizontal, and/or include advances in technology development, hence vertical, e.g. from R&D to commercialisation.
>
> (Urban *et al.*, 2015: 236)

An example of technological transfer has been the Clean Development Mechanism (CDM), which should help Annex 1 countries to reduce greenhouse gas reduction targets through investment in non-Annex 1 countries. While it was not primarily designed to transfer technology to developing countries, there was nevertheless hope that through the CDM "industrialized countries would meet some of their emissions targets through the transfer of new, clean technologies to developing" (Ockwell and Mallett, 2013: 115). However, "the CDM has been criticized for failing to achieve either sufficient technology transfer or the development dividend" (Forsyth, 2007: 1686). Moreover, Pier Paolo Saviotti highlights that transferring technology is far from an easy policy endeavour and that the interplay between private and public organisations are highly complex. Technological development and innovation are complex, messy and non-linear, and are part of broader economic, social and political development (Saviotti, 2005). One can see that transferring a technology from one socio-technical regime to another will therefore face not just technological but also social, political and economic challenges.

It is important to acknowledge that "not all low-carbon development is pro-poor, and some options offer far better benefits for the poor than others" (DIIS, 2009: 1). This is inherent within the whole discourse of technology transfer, as it is not completely clear that increasing the transfer of low-carbon technology towards low- and middle-income countries will actually benefit the poor countries or even LDCs. There is an assumption that by accessing low-carbon technology, poorer countries will "skip certain 'dirty' stages associated with development" (Ockwell and Mallett, 2013: 111). In such a discourse, poverty reduction is just a positive side effect of economic development. One can also imagine that low-carbon technologies could lead to rising inequalities, as access to these technologies or access to profit from economic growth could be concentrated in a few hands or a few countries.

However, the CDM that was supposed to transfer technologies to developing countries has been focused on middle-income countries: "less than one percent of all CDM projects to date have been implemented in LDCs (42 out of 4,660 projects)" (Funder *et al.*, 2009: 20). Rob Byrne and colleagues suggest that 83 per cent of total investment has gone to Brazil, Russia, India and China (Byrne *et al.*, 2012). On the other hand, LDCs have contributed only about 0.2 per cent of the certified emissions reductions (De Lopez *et al.*, 2009).

David Ockwell and Alexandra Mallett argue that if countries do not have sufficient capacity to absorb low-carbon innovation. "they are also unlikely to be able to adapt and incrementally change these technologies (both processes of innovation) in order to sustain low carbon development trajectories in the long term" (Ockwell and Mallett, 2013: 110). This could be seen as an explanation why significant proportion of low-carbon technologies and funding have been attracted by the BRIC countries (Brazil, Russia, India and China). UNCTAD argues that "unless the LDCs adopt policies to stimulate technological catch-up with the rest of the world, they will continue to fall behind other countries technologically and face deepening marginalisation in the global economy" (UNCTAD, 2007: i).

Scholars such as Byrne and colleagues (2012) and Markus Lederer (2010) have pointed out that most CDM projects are large-scale, centralised energy options. One could argue that smaller-scale decentralised approaches would be more adequate for poor communities. One of the dangers is that most low-carbon technologies have not served the purpose of poverty reduction but in fact helped to implement large-scale projects that benefit large-scale actors such as international corporations.

The case of the UNCHR Solar Cooker Project in Darfur

The armed conflict in Darfur, Sudan is estimated to have killed 300,000 people and resulted in more than two million displaced refugees. Many have fled to refugee camps in neighbouring Chad and the Central African Republic (Save Darfur, 2015). Life in these refugee camps is characterised by poverty and hardship. Women and girls are especially vulnerable, partly due to the traditional roles that they play in African communities. For example, women and girls are responsible for collecting fuelwood for cooking and heating:

> Fuelwood collection is traditionally thought of as a 'women's task', since it is a part of the cooking process. Rarely is cooking fuel provided by the humanitarian community, and even more rarely do men collect the wood. The burdens associated with collecting fuel fall almost exclusively on women and girls.
>
> (WCRWC, 2006a: 1, cited in Urban and Lind, 2011: 28)

This means that women and girls have to leave the relative security of their camps and venture out into foreign territory to collect fuelwood. As the Women's Commission for Refugee Women and Children (WCRWC) reports, this exposes them to risks of "rape, assault, abduction, theft, exploitation or even murder" (WCRWC, 2006a: 1). The WCRWC therefore "advocates for alternative energy options to reduce the vulnerability of women and girls associated with fuelwood collection (WCRWC, 2006a, 2006b)" (Urban and Lind, 2011: 28).

The Solar Cooker Project was driven by a consortium of NGOs, made up of TchadSolaire, Christian Outreach Relief Development (CORD), Solar Cookers International, the Dutch aid organisation KoZon and the NGO Jewish World Watch (JWW), operating in refugee camps run by the United Nations High Commission for Refugees (UNHCR). CARE International was initially also involved. This consortium provided solar cookers to Darfuri refugee women and girls in refugee camps in Chad as an alternative to using fuelwood. This initiative showed how renewable energy can contribute to wider humanitarian goals such as improving people's lives, security and gender equality and reducing pressure on the environment. The initiative initially supplied solar cookers to the Iridimi and Ouré Cassoni refugee camps in Chad, supplying about 7,000 people, a proportion of the population living in the camps. In 2006, JWW led the Solar Cooker Project and expanded the initiative's reach to all refugees in the Touloum, Iridimi and Ouré Cassoni camps, thereby helping up to 60,000 people (Urban and Lind, 2011). By 2011, the consortium had expanded their activities to another refugee camp in Farchana, thereby helping nearly 100,000 people (JWW, 2011).

> Each family in the refugee camp gets two solar cookers distributed by the NGO for free. Every women and girl of the age of 15 or older gets trained how to use and produce a solar cooker. The 'CooKit' cookers

> are made of cardboard and foil and are manufactured locally in the refugee camps by the women refugees.
>
> (Urban and Lind, 2011: 28–9)

One solar cooker was for cooking sorghum, the stable crop, and the other was for cooking a sauce or vegetables (JWW, 2011). The results were positive, with about half of the women and girls not collecting any fuelwood at all since taking part in the initiative, while a quarter collected fuelwood once a week, in comparison to up to every day before the initiative began (JWW, 2007). It is reported that two solar cookers are estimated to save one ton of wood annually (JWW, 2011). A few drawbacks exist, such as longer cooking times, the lower lifespan of the cookers requiring replacement every few months and the need to accept solar cooking as an alternative to traditional cooking methods (Urban and Lind, 2011).

Most strikingly, however, the Darfur Solar Cooker Project was discontinued in 2015 as food insecurity is the most immediate pressing security risk for the Darfuri refugees. Due to severe under-funding, food rations have been cut by the World Food Program (WFP) and the UNHCR. It is estimated that Darfuri refugees are currently living on fewer than 850 calories per person per day; some are even reported to be receiving as few as 250–500 calories per person per day. This is far lower than the WFP-recommended daily intake of 2,100 calories per person per day and is posing a severe threat to the health, well-being and survival of refugees (JWW, 2015). The pressure on food security and nutrition means that donors and aid organisations, such as those running the Darfur Solar Cooker Project, have moved away from focusing on humanitarian aid through renewable energy to increasing food supplies to avoid starvation. Obviously, the Solar Cooker Project can only help refugees if there is food available to cook. The situation is so dire that gender empowerment, reducing sexual harassment, increasing security, providing modern energy access and environmental sustainability all have to take a back seat as humanitarian aid focuses on the very basics: providing food rations for starving refugees.

It is reported that the UNHCR and other UN humanitarian agencies may be on the brink of bankruptcy and unable to meet the basic needs of millions of people due to the sheer scale of the refugee crisis in the Middle East, Africa and Europe, as well as due to shortfalls in income. The UNHCR is reporting a cut in funding of 10 per cent in 2015 compared to the previous year, with the refugee budget for Syria being only about 40 per cent of the budgeted requirements in 2015. Similarly, the WFP is under-funded by more than 60 per cent in 2015, according to senior figures within the UN (Erlanger and De Freytas-Tamura, 2015).

Table 7.2 The Darfur Solar Cooker Project

Substantive dimension	*Procedural dimension*	*Social policy dimension*
About 100,000 people in the Darfuri refugee camps in Touloum, Iridimi, Ouré Cassoni and Farchana in Chad benefited from solar cookers. Women and girls had alternative means of cooking and did not need to leave the refugee camps to collect fuelwood outside. This meant that they were less at risk of rape, assault, abduction, theft, exploitation and even murder. At the same time, women and girls received training and developed skills to build the solar cookers, they received income for doing so and they had access to modern energy. From an environmental perspective, this contributed to climate change mitigation and may have helped to preserve woods and forests that may otherwise have been in decline due to the need for fuelwood.	Procedural aspects of the Solar Cooker Project and other similar initiatives, such as governance and accountability, depend mainly upon the aid organisation. They need to be accountable and transparent to donors, such as governments, firms or individuals. At the same time, humanitarian aid programmes remain somewhat ad hoc as they depend upon the changing circumstances of the refugees and the conflict zones. For the Solar Cooker Project, this meant that the project was shelved after about a decade due to more pressing needs for food security.	Sudan and South Sudan are considered as fragile (or failed) states. Chad, where the mentioned Darfuri refugee camps are located, is also a fragile state. This means that the state's central government is weak and has little control over parts of its territory, there is no provision of public services, there is widespread corruption, criminality and economic destitution and there are a large number of refugees. Such states can therefore offer only very little to the social security of refugees.

Substantive dimension

In terms of the substantive dimension of social and environmental justice, the Darfur Solar Cooker Project offered clean energy access to about 100,000 people in refugee camps in Chad. Women and girls thereby received alternative means of cooking and did not need to leave the refugee camps to collect fuelwood outside. This meant that they were less at risk of rape, assault, abduction, theft, exploitation and even murder. As many of the women and girls are illiterate and have lost access to their land, they are unable to sustain their former livelihoods in agriculture. Through the Solar Cooker Project, women and girls received training and developed skills to build the solar cookers, they received income for doing so and they had access to modern energy. While this was not formal employment and incomes were moderate, it nevertheless provided an opportunity for women and girls to generate their own income, to be more independent and to learn a new set of skills and knowledge. This is reported to have given women and girls a sense of pride as they are contributing to their household and its finances (Urban and Lind, 2011).

The WCRWC reports that conflicts between local people from Chad and Darfuri refugees were primarily caused by competition over fuelwood in recent years, as local supplies have dwindled and market prices have increased (WCRWC, 2006b). One of the reasons why refugee women and girls are attacked by locals in Chad is because of the lack of sufficient fuelwood resources and the perceived intrusion of foreigners cutting down the trees that locals need for fuelwood. The project staff reported a positive environmental and conflict-related impact of the Solar Cooker Project. Not only were trees and woodlands preserved by the Solar Cooker Project, but there was the potential that the project may have contributed to peace-building and stability in the region.

From an environmental perspective, the Solar Cooker Project contributed to climate change mitigation and may have helped to counter deforestation and desertification, preserving woodlands that may otherwise have been in decline due to the need for fuelwood.

Procedural dimension

The procedural dimension of social and environmental justice for the Darfur Solar Cooker Project depended mainly upon the aid organisations implementing and managing the project. They needed to be accountable and transparent to donors, such as governments, firms or individuals.

At the same time, humanitarian aid programmes remain somewhat ad hoc as they depend upon the changing circumstances of the refugees and the conflict zones. For the Solar Cooker Project, this meant that the project was shelved after about a decade due to more pressing needs for food security. While projects such as the Solar Cooker Project have the potential to transform the lives and livelihoods of tens of thousands of people for the better, they are unfortunately linked to wider political and economic agendas. The global political economy of this specific project meant that rapidly rising numbers of refugees in the Middle East, North Africa and Europe and limited budgets put a strain on the ability of donor agencies such as the UNHCR to help locally in Darfuri refugee camps. This meant that food rations had to be cut, not only in refugee camps in Chad but worldwide, and as a consequence JWW, the aid organisation that was running the Solar Cooker Project, decided to stop the initiative and instead invest its resources in increasing food rations to refugees to avoid starvation.

This highlights three main issues:

1 For refugees who are struggling to survive, there are far more pressing issues than access to clean, modern energy and even more pressing

issues than keeping women and girls safe from sexual violence. Providing sufficient food and water to impoverished refugees is clearly the number one priority for protecting human lives.

2 While projects such as the Darfur Solar Cooker Project are making important contributions to improving the lives of people and contributing to gender empowerment, security and stability, income generation, training and environmental sustainability, they may appear to be like a drop in the ocean. The sheer scale of human misery across the globe caused by war, armed conflict, repressive regimes and natural disasters such as floods, droughts and tropical storms (which are often exacerbated by climate change) poses huge challenges to humanitarian aid and emergency responses.

3 Global problems require global solutions. The issues addressed by the Darfur Solar Cooker Project are not limited to the mentioned refugee camps in Chad. Many of the problems faced by refugees, particularly women and children, are similar or even the same in many parts of the world. Women and children are especially exposed and vulnerable while fleeing their homes and adversely affected by the harsh conditions in refugee camps. Sexual violence is not limited to fuelwood collection. Against this backdrop, the Darfur Solar Cooker Project would have ideally sustained sufficient financial and logistic support alongside support for improving the food rations of refugees. Ideally, the refugees in Chad and elsewhere around the world would not have to depend upon the important and well-meaning, but short-term, fragmented and often ad hoc aid interventions of charities. Instead, a well-funded globalised intervention coordinated by UN agencies would be able to ensure food security, as well as safety and access to clean water and modern energy in refugee camps around the world. While this sounds simple, the reality is currently far from it.

Social implications, challenges and solutions

Providing modern energy access, for example through solar cookers, to the world's poorest based in refugee camps can have positive social implications, including co-benefits for gender empowerment, increased security and income generation. The Darfur Solar Cooker Project benefited up to 100,000 people in refugee camps in Chad (JWW, 2011). The project's main aim was to increase the security of women and girls and to reduce sexual violence, by providing alternatives to leaving the refugee camps to collect fuelwood. This was achieved, with about half of the women and girls not collecting any fuelwood at all since taking part in the initiative, while a quarter collected fuelwood once a week, in comparison to up to every day before the initiative started (JWW, 2007). It provided opportunities for training and income generation for women and girls who built and assembled the solar cookers. At the same time, it enabled the transfer of

Table 7.3 Social implications, challenges and solutions

Social implications	*Challenges*	*Solutions*
Providing modern energy access, for example through solar cookers, to the world's poorest based in refugee camps can have positive social implications, including co-benefits for gender empowerment, increased security and income generation.	While governments, donors and firms can provide some relief, there is a lack of consistent formal social policies to lighten the burden of the poorest in refugee camps. The reliance upon informal welfare (such as donations) and informal employment and training (such as NGOs training women and girls to produce solar cookers) poses a challenge due to the often temporary, ad hoc and fragmented nature of these emergency relief activities. There is a constant risk that more important priorities crop up and push the original relief activities aside. Another challenge is the constant need for fund-raising and a long-term lack of stable funding.	Governments from developed countries need to invest more in the welfare and well-being of refugees to meet basic human needs, and should consider providing modern energy access. Ideally, funding should be centrally raised and distributed worldwide through UN agencies such as the UNHCR. The lack of formal social policies in recipient countries and communities may partly be counteracted by a more consistent, better funded global approach.

low-carbon energy technology. As the technology was rather low-tech, constructed of cardboard and foil, it was easy to assemble and could be built even in the most remote and basic environments. In this specific case, both 'hardware' and 'software' – namely, skills, expertise and knowledge – were transferred to the recipients. Very rarely are poor refugees the recipients of low-carbon technology transfer. One can therefore only congratulate the project on achieving multiple goals of a social, environmental and technical nature.

In light of the current refugee crisis, it is unlikely that projects such as the Darfur Solar Cooker Project will be revived, as more pressing issues are putting a strain on scarce resources. It is therefore regrettable, although completely understandable, that the project was recently discontinued due to a need to prioritise other emergency relief activities – in this case, supporting the food security of the refugees as global food rations in refugee camps were cut due to under-funding of UN agencies.

While governments, donors and firms can provide some relief, there is a lack of consistent formal social policies to lighten the burden on the poorest in refugee camps. The reliance upon informal welfare (such as donations) and informal employment and training (such as NGOs training women and girls to produce solar cookers) poses a challenge due to the often temporary, ad hoc and fragmented nature of these emergency relief activities. There is a constant risk that more important priorities will crop up and push the original relief activities aside. Another challenge is the constant need for fund-raising and a long-term lack of stable funding.

Governments from developed countries need to invest more in the welfare and well-being of refugees to meet basic human needs, and by doing so should consider providing modern energy access. Ideally, funding should be centrally raised and distributed worldwide through UN agencies such as the UNHCR. The lack of formal social policies in recipient countries and communities may partly be counteracted by a more consistent, better funded global approach. Hence, NGOs and to some extent also the private sector are vital partners for emergency relief. The private sector's potentially important role can be seen in initiatives such as those of the IKEA Foundation (the philanthropic organisation of the global furniture company). For instance, it has teamed up with the UNHCR to produce, test and supply solar-powered flat-pack shelters for refugees. The shelters provide accommodation and privacy for refugees, while at the same time providing lighting after dark (UNHCR, 2015a; Zimmer, 2013). The UNHCR and the IKEA Foundation also run the Brighter Lives for Refugees campaign, which provides solar lamps for refugee camps, for example in Ethiopia and for Syrian refugee camps in Jordan (UNHCR, 2015b). Other options might be to increase technology transfer under the CDM for low-income countries, especially targeting transfer of low-carbon energy technology to refugee camps.

While the private sector might be ramping up its investments in this field, UN agencies are faced by severe under-funding due to the rapidly increasing number of people displaced by conflicts in the Middle East and in Africa, as well as a lack of global commitment to providing more funding for the world's poorest. The year 2015 was marked by an unprecedented number of people fleeing conflicts around the world, mainly from Syria, but also Afghanistan, Iraq, Pakistan, Eritrea and Nigeria as well as an increasing number of people fleeing violence and conflict in Latin America, making their way to the US. EU leaders are pledging support for refugees in the Middle East, while at the same time struggling to deal with the refugees who have arrived in the EU. On a positive note, the global community is more aware than before of the issues that refugees are facing and acts of solidarity and humanity can be witnessed in many places. However, there is a risk that less visible conflicts, such as the one in Darfur, are being forgotten and sidelined. This crisis calls for increased investments to support refugees, both those in refugee camps near to conflict areas and those seeking asylum in OECD countries, as well as for increased transfer of low-carbon energy technologies to the world's poorest. It also calls for coordinated governments' actions in addition to

NGO and private-sector action, but this situation should also encourage each and every person in the richer countries to help individually, such as by donating to the world's poorest.

This chapter has discussed how providing solar energy products such as solar cookers to refugees could be a start to making low-carbon technology transfer more socially just and more accessible to the world's poorest and disadvantaged. At the same time, the chapter has evaluated these issues critically, particularly due to the temporary, fragmented, ad hoc and under-funded nature of many emergency relief activities led by donors and NGOs. Low-carbon energy access for people in refugee camps has the potential to impact positively on their lives and livelihoods while at the same time contributing to climate-friendly development and wider environmental sustainability. Yet, for this to happen at a larger scale and over longer time frames requires more commitment to funding, particularly by the governments of developed countries.

References

Bell, M. (2009), "Innovation capabilities and directions of development", STEPS Working Paper 33, Brighton: STEPS Centre.

Bell, M. and Pavitt, K. (1993), "Technological accumulation and industrial growth: contrasts between developed and developing countries", *Industrial and Corporate Change*, 2(2), 157–210.

Biswas, W. K., Bryce, P. and Diesendorf, M. (2001), "Model for empowering rural poor through renewable energy in Bangladesh", *Environmental Science and Policy*, 4(6), 333–44.

Byrne, R., Smith, A., Watson, J. and Ockwell, D. (2011), "Energy pathways in low carbon development: from technology transfer to socio-technical transformation", STEPS Working Paper, Brighton: STEPS Centre, available at: http://steps-centre.org/publication/energy-pathways-in-low-carbon-development-from-technology-transfer-to-socio-technical-transformation-2/ (accessed 27 March 2016).

Byrne, R., Smith, A., Watson, J. and Ockwell, D. (2012), "Energy pathways in low-carbon development: the needs to go beyond technology transfer", in D. Ockwell and A. Mallett (eds), *Low-Carbon Technology Transfer: From Rhetoric to Reality*, London: Routledge, pp. 123–42.

Cohen, W. and Levinthal, D. (1990), "Absorptive capacity: a new perspective on learning and innovation", *Administrative Science Quarterly*, 35(1), 128–52.

De Lopez, T., Ponlok, T., Iyadomi, K., Santos, S. and McIntosh, B. (2009), "Clean Development Mechanism and Least Developed Countries: changing the rules for greater participation", *The Journal of Environment and Development*, 18(4), 436–52.

DFID (Department for International Development) (2002), *Energy for the Poor: Underpinning the Millennium Development Goals*, London: DFID, available at: www.ecn.nl/fileadmin/ecn/units/bs/JEPP/energyforthepoor.pdf (accessed 27 March 2016).

DIIS (Danish Institute for International Studies) (2009), *Reducing Poverty Through Low Carbon Development. Recommendations for Development Cooperation in*

Least Developed Countries, Copenhagen: DIIS, available at: www.ciaonet.org/attachments/15363/uploads (accessed 27 March 2016).

Erlanger, S. and De Freytas-Tamura, K. (2015), "UN funding shortfalls and cuts in refugee aid fuel exodus to Europe", *New York Times*, 19 September, available at www.nytimes.com/2015/09/20/world/un-funding-shortfalls-and-cuts-in-refugee-aid-fuel-exodus-to-europe.html?_r=0 (accessed 27 March 2016).

Forsyth, T. (1999), *International Investment and Climate Change: Energy Technologies for Developing Countries*, London: Earthscan.

Forsyth, T. (2007), "Promoting the 'development dividend' of climate technology transfer: can cross-sector partnerships help?", *World Development*, 35(10), 1684–98.

Funder, M., Fjalland, J., Ravnborg, H. M. and Egelyng, H. (2009), *Low Carbon Development and Poverty Alleviation: Options for Development Cooperation in Energy, Agriculture and Forestry*, DIIS Report 2009:20, Copenhagen: DIIS, available at: http://subweb.diis.dk/graphics/Publications/Reports2009/DIIS_Report_2009-20_Low_Carbon_Development_and_Poverty_Alleviation.pdf (accessed 27 March 2016).

Hedger, M. M., Martinot, E., Onchan, T., Ahuja, D., Chantanakome, W., Grubb, M., Gupta, J., Heller, T. C., Junfeng, L., Mansley, M., Mehl, C., Natarajan, B., Panayotou, T., Turkson, J. and Wallace, D. (2000), "Enabling environments for technology transfer", in B. Metz, O. R. Davidson, J.-W. Martens, S. N. M. van Rooijen and L. Van Wie McGrory (eds), *Methodological and Technological Issues in Technology Transfer: A Special Report of IPCC Working Group III*, Cambridge: Cambridge University Press, pp. 105–41.

IEA (International Energy Agency) (2011), *World Energy Outlook 2011*, Paris: OECD/IEA.

IPCC (Intergovernmental Panel on Climate Change) (2007), *Climate Change 2007: Synthesis Report*, Fourth Assessment Report of the Intergovernmental Panel on Climate Change, Cambridge: Cambridge University Press.

IPCC (Intergovernmental Panel on Climate Change) (2014), *Climate Change 2013: Synthesis Report*, Fifth Assessment Report of the Intergovernmental Panel on Climate Change, Cambridge; Cambridge University Press.

Jakob, M. and Steckel, J. C. (2013), "How climate change mitigation could harm development in poor countries", *WIREs Climate Change*, 5(2), 161–8.

JWW (Jewish World Watch) (2007), *Solar Cooker Project Evaluation*, Encino: JWW.

JWW (Jewish World Watch) (2011), "Help the women of Darfur", factsheet, Encino: JWW, available at: www.jewishworldwatch.org/media/scp/JWW_FactSheet_Mar11_v3.pdf (accessed 27 March 2016).

JWW (Jewish World Watch) (2015), "Solar Cooker Project: food crisis in the Darfuri refugee camps", Encino: JWW, available at: www.jww.org/projects/ontheground/solar-cooker-project-redirect (accessed 8 July 2016).

Juma, C. and Yee-Cheong, L. (eds) (2005), *Innovation: Applying Knowledge in Development*, Task Force on Science, Technology and Innovation, United Nations Millennium Project, New York: United Nations, available at: www.unmillenniumproject.org/documents/Science-complete.pdf (accessed 27 March 2016).

Lederer, M. (2010), "Evaluating carbon governance: the Clean Development Mechanism from an emerging economy perspective", *The Journal of Energy Markets*, 3(2), 1–24.

Lockwood, M. (2013), "The political economy of low carbon development", in F. Urban and J. Nordensvärd (eds), *Low Carbon Development: Key Issues*, London: Earthscan, pp. 25–38.

Modi, V., McDade, S., Lallement, D. and Saghir, J. (2005), *Energy Services for the Millennium Development Goals*, Energy Sector Management Assistance Programme, United Nations Development Programme, New York: UN Millennium Project and World Bank, available at: www.unmillenniumproject.org/documents/MP_Energy_Low_Res.pdf (accessed 27 March 2016).

Ockwell, D. and Mallett, A. (eds) (2012), *Low Carbon Technology Transfer: From Rhetoric to Reality*, London: Routledge.

Ockwell, D. and Mallett, A. (2013), "Low carbon innovation and technology transfer", in F. Urban and J. Nordensvärd (eds), *Low Carbon Development: Key Issues*, London: Earthscan, pp. 109–29.

Ockwell, D., Watson, J., MacKerron, G., Pal, P., Yamin, F., Vasudevan, N. and Mohanty, P. (2007), *UK–India Collaboration to Identify the Barriers to the Transfer of Low Carbon Energy Technology*, London: DEFRA.

Ockwell, D., Mallett, A., Haum, R. and Watson, J. (2010), "Intellectual property rights and low carbon technology transfer: the two polarities of diffusion and development", *Global Environmental Change*, 20(4), 729–38.

Save Darfur (2015), "Darfur update", available at: http://savedarfur.org/ (accessed 27 March 2016).

Saviotti, P. (2005), "On the co-evolution of technologies and institutions", in M. Weber and J. Hemmelskamp (eds), *Towards Environment Innovation Systems*, Berlin: Springer, 9–31.

UNCTAD (United Nations Conference on Trade and Development) (2007), *The Least Developed Country Report 2007*, Geneva: United Nations.

UNFCCC (United Nations Framework Convention on Climate Change) (1992) *The United Nations Framework Convention on Climate Change*, Bonn: UNFCCC, available at: http://unfccc.int/resource/docs/convkp/conveng.pdf (accessed 18 October 2012).

UNHCR (United Nations High Commission for Refugees) (2015a), "A safe place to call home", available at: www.unhcr.org/pages/52a5c44f6.html (accessed 27 March 2016).

UNHCR (United Nations High Commission for Refugees) (2015b), "The Brighter Lives for Refugees campaign", www.unhcr.org/brighterlives/ (accessed 27 March 2016).

Urban, F. and Lind, J. (2011), "Low carbon energy and conflict: a new agenda", *Boiling Point*, 59, available at: www.hedon.info/BP59_Low+carbon+energy+and+conflict (accessed 27 March 2016).

Urban, F., Siciliano, G., Sour, K., Lonn, P. D., Tan-Mullins, M. and Mang, G. (2015), "South–South technology transfer of low carbon innovation: Chinese large hydropower dams in Cambodia", *Sustainable Development*, 23(7–8), 232–44.

Urban, F., Ting, M. B. and Lakew, H. (2013), "Poverty reduction and economic growth in a carbon constrained world: the case of Ethiopia", in F. Urban and J. Nordensvärd (eds), *Low Carbon Development: Key Issues*, London: Earthscan, pp. 228–39.

WCRWC (Women's Commission for Refugee Women and Children) (2006a), *Finding Trees in the Desert: Firewood Collection and Alternatives in Darfur*, New York: WCRWC.

WCRWC (Women's Commission for Refugee Women and Children) (2006b), *Beyond Firewood: Fuel Alternatives and Protection Strategies for Displaced Women and Girls*, New York: WCRWC, available at: womensrefugeecommission.org/resources/document/165-beyond-firewood-fuel-alternatives-and-protection-strategies-for-displaced-women-and-girls (accessed 27 March 2016).

WRI (World Resources Institute) (2005), *Navigating the Numbers: Greenhouse Gas Data and International Climate Policy*, chapter on cumulative emissions. http://pdf.wri.org/navigating_numbers.pdf (accessed 8 June 2016).

Zimmer, L. (2013), "IKEA unveils solar-powered flat pack shelters for easily deployable emergency housing", *Inhabitat*, 11 November, available at: http://inhabitat.com/ikeas-solar-powered-flat-pack-refugee-shelters-offer-easily-deployable-emergency-housing/ (accessed 27 March 2016).

Conclusion and discussion

Could a global low-carbon future be socially just?

Johan Nordensvärd

Can low-carbon development be socially just?

The argument of this book is that low-carbon development is essential for mitigating climate change and enabling development in a carbon-constrained world, but there are risks that low-carbon development might come at a cost that is both social and economic. These risks need to be carefully assessed and reduced. The main aim of the book was to explore, critically analyse and propose different ways in which low-carbon development could be socially viable in both developed and developing countries. The different case studies show that low-carbon development is no magic bullet and that the meaning of the concept is ambiguous. Low-carbon development is often equated with low-carbon growth, which in itself creates two problems:

1. a lack of focus on human development, social sustainability and environmental justice;
2. an overly narrow focus on carbon emission reductions while sustaining economic growth which historically has been difficult to achieve.

Even though the mainstream discourse promotes the notion that achieving low-carbon growth is possible and desirable, it should not be disregarded that low-carbon development might in some cases imply the opposite, namely discussing options for:

- no growth or even degrowth scenarios;
- more socially just distribution of resources on a global scale.

The social dimension is not only neglected within the sustainability framework but plays a very small role within the low-carbon development discourse. It is often assumed that low-carbon development will lead to sustainable economic growth and that its benefits will trickle down to the poorest. The issue is that benefits are easier to track for vested interests than for the rest of society. This book shows that low-carbon development faces challenges in appeasing vested interests on the one hand and mitigating some of the social costs stemming from these policies on the other.

Low-carbon development for whom?

The case study on the social implications of the large hydropower Kamchay Dam in Cambodia (Chapter 3) showed that large hydropower dam projects are complex and have large social impacts. The Kamchay Dam was built by the Chinese state-owned enterprise (SOE) Sinohydro and financed by the Chinese-owned ExIm Bank. The dams industry considers hydropower dams as a source of climate-friendly low-carbon energy which can provide cheap electricity, including for poor people in developing countries. Often it is argued that this investment will help both the country and the people to be lifted out of poverty, but in reality the situation is more complex. The positives should be clear, as electricity is seen as a prerequisite for development. Most of the electricity that has been generated by the Kamchay Dam, Cambodia's first large dam, is not used by the local population but in the capital Phnom Penh. The direct benefits for local people are limited while at the same time they have been negatively affected by declines in livelihoods, lack of access to natural resources and flooding of their homes and land (in September 2015). This unequal balance of benefits and costs is still not adequately considered in many low-carbon development projects. The overarching lesson learned from the book's case studies is the need to consider the question of who will benefit from low-carbon development. The discussion on social sustainability and environmental justice has attempted to highlight these issues; similar to those affected by climatic impacts, some of the most adversely affected people are not the primary beneficiaries of any large-scale projects, but the poor and vulnerable. In the Kamchay Dam case in Cambodia, the half-hearted implementation of social protection illustrates a larger global problem. There are global standards and codes of conduct for environmentally and socially sensitive infrastructure projects such as hydropower dams but at the moment these standards and codes of conduct are not globally legally binding. It is down to national governments and global corporations to regulate these matters. The situation is slowly improving, for example through the work of benchmarking firms such as International Rivers, who produced a benchmarking report for large dam-builders in China, the world's largest dam-building nation (International Rivers, 2015). This could be extended and applied to dam-building firms worldwide. Standards formulated by the World Bank/IFC, the IHA's Hydropower Sustainability Assessment Protocol, the recommendations of the World Commission on Dams and the Equator Principles for financial institutions are already in place. The problem is that these standards and codes of conduct are not legally binding at present and depend upon the voluntary actions of dam-builders. In addition to making these legally binding, it would be useful for future dam contracts to be regulated by an international governing body to prevent severe social implications for the local population. This is a massive governance problem, as it is unclear who should enforce these codes of conduct and what consequences there should be if they are breached. The standards of the dam-builder Sinohydro are, according to the International Rivers benchmarking report, supposed to be the best of all Chinese dam-builders (International Rivers, 2015); yet the Kamchay Dam was built in a national park and had negative social implications for the local population (Urban *et al.*, 2015).

The governance situation does not make it easier when it comes to the complexity of the contracts that involve host countries and dam-builders. One must distinguish between the different roles that institutions have in the process of dam-building. There are financiers/investors, developers, builders and contractors, component suppliers and dam operators (Urban *et al.*, 2013):

- *Financiers/investors* include the World Bank, the Asian Development Bank and the ExIm Bank, which as a funder of large dams could be considered equal to the World Bank, which also funds large dams. *Financiers/investors* may not have much power over social issues such as resettlements and compensation for affected people, but they could refrain from investing in morally and socially irresponsible projects. They are usually also bound by the Equator Principles for investors.
- *Developers* may be influenced by the host government, for example with regard to site selection. However, they may be involved in initiating surveys and planning the dam, including its design and construction, which often includes major decisions about the size of the dam, its generating capacity, the size of the reservoir that will flood the area, and so forth. This highlights a complex interaction between countries and developers and builders in which the affected people often have little say. There is also no global code of conduct to decide who is ultimately responsible for the dam's impacts. Depending upon the type of contract, and particularly for Build, Operate, Transfer (BOT) contracts, this may also include decisions about compensation payments, including how much money should be made available. Developers should therefore be well aware of the full extent of the dam-building process and its impacts and able to make informed decisions about its sustainability and whether or not a project should go ahead.
- The *builders* of the dam or the *contractors* who either build the dam entirely or supply components (such as the turbines) are in charge of the engineering and the actual construction and implementation of the dam, sometimes including connecting it to the grid. While they may not be aware of the full extent of the socially and environmentally adverse effects of the dam, they are usually sufficiently informed to decide whether or not the project is morally and socially defensible.

- *Dam operators* are those who are in charge of managing the dam on a daily basis. Like the builders and contractors, they may not be aware of the full extent of the socially and environmentally adverse effects of the dam, but they are usually sufficiently informed to decide whether or not the project is morally and socially defensible.

Sometimes these roles are divided between different actors; sometimes one actor has multiple roles, such as being developer, builder and dam operator all in one, as was the case with Sinohydro's/PowerChina Resources' role for the Kamchay Dam (Urban *et al.*, 2015).

The findings of the Kamchay Dam case study suggest that the degree to which Chinese dam builders and financiers tend to follow social and environmental standards and best practice for planning and implementing hydropower dams is very much debatable. However, the national government also plays a large role in choosing sites for the dams and setting social standards. Sometimes national governments are responsible for (forced) resettlement and compensation to the affected local population. The pressure should therefore not only be on dam-building firms and financiers to minimise the social and environmental impact of large dams, mining operations and large infrastructure projects; host governments should also be held accountable. This will prove to be a challenge for many developing countries, as these projects are tied to lucrative investment contracts. It is as naive to believe that governments always act in the interests of the affected population as it is to believe that international corporations have an intrinsic interest in a socially just outcome for the affected population. This challenge becomes even more severe in the case of poor countries such as Cambodia, which lack a comprehensive welfare state and where most affected people are dependent upon informal social welfare to cope. When communities risk losing their livelihoods, they will have difficulties coping when the government and corporations do not adequately support them. A superficial solution would be to improve the contracts between investors, dam-builders and the host country to include more extensive social policy/social protection, but in the end it will be a matter of global regulation for investment to include social policy and social and environmental sustainability at a minimal level for all international hydropower dam projects.

The main issue here is to understand that while approaches to low-carbon development and investments in low-carbon energy have positive goals, such as mitigating carbon emissions, potentially providing energy access and potentially contributing to national economic growth, the adverse effects may be felt by the poorest and most vulnerable, while the benefits may be enjoyed by the elites and the better-off. There is therefore a need for more comprehensive, transparent regulation and monitoring of large-scale low-carbon energy projects, such as large dams, and their social implications.

Low-carbon governance

The case study on Plantar's large-scale industrial tree plantations in Minas Gerais, Brazil (Chapter 5) showed that the environmental and social implications of an accepted CDM project are not straightforward and remain a highly contentious issue. The business of buying and selling offset carbon credits is not only lucrative but also open to corruption and fraudulent behaviour. These market solutions may open up grey areas that cast doubt on both the procedural and substantive dimensions of low-carbon growth. Brazil, for example, has become a top seller of carbon offsets. Proponents argue that setting a price and trading carbon dioxide equivalent (CO_2e) emissions is a cost-effective way to reduce pollution.

The Plantar case study examined a particularly difficult project where the environmental and social aspects are not clear-cut. It seems counterproductive that offsetting credits should be given to monocultural non-native eucalyptus plantations. Eucalyptus is an exogenous fast-growing tree, and eucalyptus plantations cause water stress in affected territories. The impact of water stress affects local streams and waterways upon which local communities and biodiversity depend. The native *cerrado* (savannah biome) is host to small streams, but as a result of the large-scale plantations many water springs have dried up. In addition, Plantar uses several pesticides to kill competing plants and animals, causing soil depletion and polluting land and water. Limited water supplies negatively impact basic needs, including food sovereignty. In addition, the intensive use of pesticides poisons the streams and sources of drinkable water causing local residents give up their homes and move to urban areas. Local production has been crushed by the large-scale industry. The *cerrado* was previously used for crop production, cattle, wild foods, medicines and crafts. Local food production, agriculture and stock-raising efforts depending upon biodiversity in the area have been affected, leaving many unemployed (Lohmann, 2006: 308). As a result, the local economy has shifted to a wage-based economy, forcing more people to work for the company or leave the region. Furthermore, the company plants one non-native tree species in order to fell and burn the trees in a six-year rotation cycle. There is no evidence supporting carbon neutrality claims from a short, rapid-growth life cycle (Pearce, 2002). In fact, research shows that plantations do not even begin to balance the CO_2 lost from vegetation clearance and soil disruption until anywhere from ten to 100 years in an established ecosystem.

The new climate agreement reached in Paris makes it even more important to develop a new form of low-carbon governance that is accountable, environmentally just and socially sustainable. The Plantar case study showed that there is some way to go before we get there. Emissions from a CDM project should theoretically be lower than emissions that would have happened anyway. In the CDM, additionality is assessed by a Designated Operational Entity (DOE), a private agency selected by the project participants which subsequently produces a validation report to be evaluated by the Methodology Panel of the CDM Executive Board (EB). The DOE's assessment, including baseline definition and additionality, is then presented in a Project Design Document (PDD) (UNFCCC, 2005). Reliance upon private agencies paid by the companies to furnish their evaluation studies raises doubts about the impartiality of the DOEs.

The CDM's additionality verification and guarantee process is ambiguous not only at the international level (the activities of the EB and DOEs) but also at the national level, where Designated National Authorities oversee CDM proposals. At this level, Axel Michaelowa and Katharina Michaelowa have stated, "CDM authorities do not care about additionality of CDM projects" (cited in Lohmann, 2006: 178).

The large number of CDM projects whose additionality has been questioned and contested led the EB to strengthen the verification process and provide a systematic assessment in 2007 (Gilbertson and Reyes, 2009; Michaelowa and Purohit, 2007). An increase in rejections and reviews of CDM projects resulted and by 2011 new guidelines for the establishment of performance standards were implemented (Hayashi and Michaelowa, 2013: 192) However, by 2011, many CDM projects were already coming to the end of their first term. Although several projects were reviewed, no projects that were already underway were thrown out.

After a multitude of international efforts to keep Plantar out of the CDM, in 2010 the Plantar project was finally registered and issued just over four million CERs in association with the World Bank's Prototype Carbon Fund (PCF) and the BioCarbon Fund (BioCF) for three project activities referred to as *Projeto Plantar* (Plantar Project) and managed by Plantar Carbon Ltda (Grupo Plantar, 2015). The Plantar Project was then divided into three connected CDM projects: the methane avoidance project, which was registered by the EB in 2007; the reforestation project, finally registered in 2010; and the green pig-iron production, registered in 2012. It needs to be discussed how a future with more similar offset projects can be both socially and environmentally just given that one could even debate whether the environmental rationale behind the projects are sound.

It should therefore be no surprise that the social implications are often a mere afterthought. Scholars have pointed out that extractive industries often lead to the economic and social marginalisation of residents in remote and rural areas. Often local people obtain few of the benefits that come with mining and other extractive activities on their land. It is more likely that these activities will threaten existing and viable livelihoods (Cademartori, 2002; Freudenburg, 1992). All along the production chain, workers are of course the first to be exposed to the chemical products used in the plantation process or during the burning phase. Despite the company's claims of job creation, the Curvelo Regional Labour Office (DRT) contested Plantar's working conditions and issued a summons to the company in 2002 for slave and child labour in timber extraction and charcoal production. The DRT fined Plantar after finding 194 workers without any registration in its plantation near Curvelo (Lohmann, 2006: 315). The claims were so grave that the state of Minas Gerais began to investigate the company's actions. The Minas Gerais Parliamentary Investigation Commission confirmed that in 2002 the company financed illegal outsourcing of labour. As mentioned in the official report, Plantar put workers in "precarious labour relations" and "abominable working conditions" and promoted "slave and child labour and deforestation of the *cerrado*". Problems with hygiene, housing, drinking water, food and transport have also been mentioned, which led the Commission to note the infringement of International Labour Organisation (ILO) provisions. Moreover, charcoal workers are constantly exposed to pesticides and are at a high risk of accidents.

As a result, the Plantar website now describes projects for enhanced workers safety, claiming that

> Plantar developed the Sistema de Gestão de Segurança e Saúde Ocupacional – SGSSO [Safety and Occupational Health Management System], recognizing that this is the set of best practices, which can also contribute to be the reference regarding Work Safety and Occupational Health issues.
>
> (Grupo Plantar, 2015)

However, even with this "set of best practices" in place, research in 2012 found that several Plantar workers complained about long hours, dismissal without explanation or pay, severe negative health impacts and terrible working conditions (Carbon Trade Watch and FASE-ES, 2013). There is therefore a need to rethink the governance of CDM projects and to draw upon a global framework for low-carbon governance that is socially sustainable and environmentally just.

Neo-liberal low-carbon development

In both Germany and the UK, low-carbon development is built on a neo-liberal model of governance and hereby avoids using taxation or strict regulations to reduce carbon emissions from production and consumption. Neil Brenner and Nik Theodore define neo-liberalism as an "actually existing framework of disciplinary political authority that enforces market rule over an ever wider range of social relations throughout the world economy" (Brenner and Theodore, 2002: 14). Low-carbon development has become increasingly linked with the general ideological turn towards neo-liberalism and the pursuit of market and quasi-market solutions (Pearse and Böhm, 2014). Both the German and English case studies (Chapters 4 and 6) showed how a neo-liberal design fails to include a more comprehensive social design.

The case study on the social implications of the German energy transition (*Energiewende*) and feed-in tariffs for wind energy (Chapter 4) showed that vested interests play a large role in creating a policy that is increasing costs for consumers while allowing corporate interests to profit from a low-carbon energy transition. It also depends upon how low-carbon transitions are financed and organised. The German wind energy sector has been highly dependent upon political decisions taken since the early 1990s to support a fast and ambitious expansion of wind energy. As part of a government reshuffle in late 2013 (resulting from the electoral victory of the so-called 'Grand Coalition'), the entire energy agenda was transferred to the Ministry of Economic Affairs and Energy. This includes overall energy policy, grid management and the promotion of renewable sources of energy as well as energy efficiency enhancement. Importantly, the management and reform of Germany's energy transition is thus directly linked to a comprehensive agenda of economic growth, competitiveness and innovation.

The feed-in tariffs for renewable energy (*EEG-Umlage*) are not funded by the state, nor are they subsidised through public funding or taxation; they are funded through consumer prices. Similar to the minimum wage, the government sets the absolute minimum price for electricity from renewable sources. The operators of renewable energy sites such as wind farms receive a specific sum in return for the feed-in electricity. This electricity is then sold on the stock market to generate profits. Feed-in tariffs cover the difference between revenues and costs (IWR, 2014). The major challenge is that even since feed-in tariffs for renewable energy have been capped, this has not resulted in falling prices; prices are actually rising. This can be explained by the fact that the government is granting exemption from feed-in tariffs to energy-intensive industries, such as the car industry and other large manufacturing industries, which meant a rise of costs from €1 billion to €5 billion by 2014. Industries that use large amounts of energy are only paying 0.05 ct/kWh for the feed-in tariff (virtually an exemption) compared to 6.17 ct/kWh for private households (IWR, 2014). The quota to get this reduction in energy prices has sunk from 10 GWh to 1 GWh which means that payments into the EEG account have decreased and the costs of the transition have to be shouldered by consumers (IWR, 2014a). Ironically, the way that the EEG and feed-in tariffs were reformed in 2010 has led to consumers being the economic losers while large firms, utilities and large energy providers are the economic winners of the energy/transition towards renewable energy in Germany. An increased share of wind energy and other renewables has indeed lowered energy prices, albeit only for large corporate customers and energy providers, not for individual consumers (IWR, 2014). The tendency to reduce electricity prices for German industries with high energy intensity is a truly divisive strategy as this means that households are more or less subsidising costly and maybe also unsustainable industries through their energy bills.

This could be seen as problematic from many different perspectives. First of all, one could argue that this is a form of corporate welfare whereby households more or less support corporations that use large amounts of energy. Corporate welfare is often defined as any action taken by the government that provides benefits to a corporation or industry not offered to others (Barlett and Steele, 1998), in this case the German state premier classical industrial corporations that are energy intensive. This means that other industries with lower energy use or from other countries are disadvantaged. Most critics of corporate welfare have found that there has been little corresponding return to taxpayers (Antonelli, 1995; Barlett and Steele, 1998; Moore and Stansel, 1995) and programmes favoured one company over another (Lindsey, 1992; Schatz, 1997). In this case, one could argue that energy consumers have not really benefited from the exemptions given to industries by the German government. Most important is the fact that consumers have paid for the largest part of the energy transition through rising energy costs while corporations have been reaping profits and energy-intensive corporations have received favourable financial treatment. Feed-in tariffs are in practice the preferential treatment of energy-intensive German industries that is according verdicts from EU such as EuGH v. 01.07.2014 conform with EU directives. Even if the German renewable energy policy was to create a barrier to the single market, this would be overridden by the prioritisation of climate protection and the interest in supporting renewable energy. In this case, the EU has changed attitude towards the German energy policy. The problem that the EU has is that corporations that use large amounts of energy are only paying 0.05 ct/kWh for energy compared to 6.17 ct/kWh for private households. The EU Commission argues that exempting these industries from feed-in tariffs should not be seen as a climate protection policy but actually as a subsidy and the preferential treatment of German industries (IWR, 2014).

The major social implications of German wind energy policy are that consumers have been bearing the brunt of the costs while corporations have been gaining the profits from the wind energy transition. Instead of contemplating raising money for a front-loaded investment in renewable energy, the German government focused on a model that raised the costs of consumers' energy bills without really improving their participation in renewable energy such as wind power. The lesson learned is that it is difficult to reach agreement on how a significant energy transition should be paid for and who should carry what costs. The problem is that offloading the costs onto consumers will mean that poorer households pay relatively more towards the energy transition than richer households and without a doubt more than the exempted industries. If there is no overall political solution whereby both the state and industries will pay more for the energy transition, this will lead to a destructive stalemate. There is a lack of political will to increase feed-in tariffs to pay for the lack of investments in the grid system.

The English case study (Chapter 6) also highlighted a neo-liberal policy whereby consumers and the market are supposed to pay for increasing energy efficiency in English homes. In England, given the poor quality of much of the housing stock, one effective low-carbon, anti-fuel-poverty measure is energy efficiency improvements within the homes of those considered most vulnerable to fuel poverty. Domestic energy policy has typically (at least in part) taken this approach. The case study focused on the two main energy efficiency programmes developed by the UK's Conservative–Liberal Coalition government (2010–15) and implemented in England, and critically analyses their environmental and social justice implications. The Coalition government focused on a market-based approach to energy policy generally, and to domestic energy efficiency specifically (Hamilton *et al.*, 2014; Mallaburn and Eyre, 2014).

Political rhetoric emphasised the need to reduce state spending, and policy shifted the responsibility for the funding and delivery of most measures from the state to the private sector. By the time of the General Election in May 2015, no substantive state-funded programmes remained, with responsibility for programmes being placed in the hands of energy suppliers and consumers, and changes also being made to CERT and CESP.[1] Overall, given the budget cuts, Coalition policy has resulted in substantially less funding for domestic energy efficiency schemes (Hills, 2012).

In the case of the Green Deal, for example, the whole scheme was dependent upon people being prepared to take out a loan that would be paid off through electricity bills. Firstly, the design may have acted as a deterrent as it was financed through loans that were repaid through energy bills that were bound to the property, rather than to the person who applied for the loan. This meant that households could effectively end up with a loan ascribed to them that they had not taken upon themselves. There was also no guarantee that the costs of the loan would outweigh the costs of repayment, which may have discouraged lower-income households from signing on for a significant financial risk that savings might not pay off (Richards, 2012). A second deterrent was the upfront assessment cost of up to £150 that households had to pay to be able to proceed with the process (some of which was refunded at a later date). These high fees were seen as a deterrent to lower-income households, instead favouring those with more disposable income (Vaughan, 2013). A third important deterrent for low-income households was the high interest rate that needed to be repaid: 6.92 per cent, which is higher than most mortgages. This could mean a significant cost, and the failure to repay could mean that households would be disconnected from gas and electricity supplies (The Green Age, 2015). A fourth important deterrent was the need to have a particular credit score to qualify for the deal. While there were attempts to lower the requirements, this is likely to have further limited take-up (The Green Age, 2015). These are all reasons why the most risk-averse households were not willing to sign up to this policy.

The ECO symbolises a shift from financing energy efficiency measures for the poorest through taxes to imposing this role on larger energy suppliers. The ECO requires larger energy suppliers to deliver energy efficiency measures to domestic households. One of the challenges of the ECO has been its struggle to create a uniform service. The results have been highly variable across England, and eligibility criteria differ substantially depending upon the partnership in place (a substantial change from the state-funded Warm Front Scheme, which had clear eligibility criteria and was administered at the national level). While the ECO has been successful in delivering carbon reductions, whether the housing stock is being sufficiently transformed is questionable. ECO carbon targets have been lowered, in part to reduce the burden on energy bill payers. While many have criticised the funding of the ECO as being regressive, most would argue that it is not the level of investment that is the problem but actually the source of the investment. As Mallaburn and Eyre argue:

> some ministers, regulators and public officials have periodically claimed that market forces on their own will deliver all cost-effective energy efficiency savings. This is simply not true ... these claims tend to be driven by ideological assumptions rather than by any serious examination of the scientific evidence.
>
> (Mallaburn and Eyre, 2014: 35)

Indeed, organisations such as the Association for the Conservation of Energy (ACE) and the Centre for Sustainable Energy (CSE) argue that fuel poverty dimensions of energy efficiency should be funded separately by the state to ensure the social sustainability of such schemes (ACE, 2014). One could even argue that there a particular neo-liberal path dependence for low-carbon development that leads us further down the route of green growth and carbon markets.

Low interest and insecure funding

The overall market reliance of low-carbon development has led to some social implications of how low-carbon technologies have spread at a global level. There has been a lack of interest in funding and sustaining low-carbon development projects among some of the poorest people in the world. While the technology transfer of low-carbon energy technology has reached predominantly emerging economies such as China and India, there has been little progress for poorer countries. The case study on the Darfur Solar Cooker Project (Chapter 7) focused on a project that provided solar cookers to Darfuri refugee women and girls in refugee camps in Sudan's neighbour country Chad as an alternative to using fuelwood. This is an example of how low-carbon energy can help to improve people's lives as well as reduce pressure on the environment in a context of political instability. Instead of women and girls being forced to leave the safety of their refugee camps in Chad and risk being assaulted in their pursuit of fuelwood, the solar cookers reduced (or even eliminated) the need for fuelwood and thereby allowed women and girls to stay safely in the camp (Urban and Lind, 2011). While this project and others like it are laudable, the reality is that they tend to be temporary and ad hoc, which tends to be in line with how emergency relief activities are organised.

The Darfur Solar Cooker Project was discontinued as food insecurity is the most immediate pressing security risk for the Darfuri refugees. Due to severe under-funding, food rations have been cut by the World Food Program (WFP) and the United Nations Refugee Agency (UNHCR). It is estimated that Darfuri refugees are currently living on fewer than 850 calories per person per day; some are even reported to be receiving as few as 250–500 calories per person per day. This is far lower than the WFP-recommended daily intake of 2,100 calories per person per day and is posing a severe threat to the health, well-being and survival of refugees (JWW, 2015). The pressure on food security and nutrition means that donors and aid organisations, such as those running the Darfur Solar Cooker Project, have moved away from focusing on humanitarian aid through renewable energy to increasing food supplies to avoid starvation. Obviously, the Solar Cooker Project can only help refugees if there is food available to cook. The situation is so dire that gender empowerment, reducing sexual harassment, increasing security, providing modern energy access and environmental sustainability all have to take a back seat as humanitarian aid focuses on the very basics: providing food rations for starving refugees.

It is reported that the UNHCR and other UN humanitarian agencies may be on the brink of bankruptcy and unable to meet the basic needs of millions of people as the due to the sheer scale of the refugee crisis in the Middle East, Africa and Europe, as well as due to shortfalls in income. The UNHCR is reporting a cut in funding of 10 per cent in 2015 compared to the previous year, with the refugee budget for Syria being only about 40 per cent of the budgeted requirements in 2015. Similarly, the WFP is under-funded by more than 60 per cent in 2015, according to senior figures within the UN (Erlanger and De Freytas-Tamura, 2015). While governments, donors and firms can provide some relief, there is a lack of consistent formal social policies to lighten the burden on the poorest in refugee camps. The reliance upon informal welfare (such as donations) and informal employment and training (such as NGOs training women and girls to produce solar cookers) poses a challenge due to the often temporary, ad hoc and fragmented nature of these emergency relief activities. There is a constant risk that more important priorities will crop up and push the original relief activities aside. Another challenge is the constant need for fund-raising and a long-term lack of stable funding.

The conclusion of this book is that, sadly, low-carbon development and social policy are not congruent and can involve two different sets of challenges. Low-carbon development therefore needs to go hand-in-hand with a wide-ranging transformation of the global economy and society, and needs to address wider global inequalities beyond climate change.

A broader understanding of low-carbon development

The book has touched upon some policy lessons such as implementing a more comprehensive understanding of environmentally just and socially just low-carbon development to avoid further socially and environmentally adverse effects. However, using green growth to mitigate climate change and to reduce poverty is not the only issue with regard to low-carbon development. One issue that seems to be forgotten is that the contemporary problems are not just down to a lack of technology and social organisation but also to over-consumption and over-production. Moreover, we also tend to forget that climate change is just one environmental problem that arises from over-consumption and over-production, while other environmental problems tend to be neglected.

For low-carbon development and climate change mitigation, the common approach is often too narrow as it tends to neglect other social and environmental implications that go beyond greenhouse gas emissions, such as air pollution, water pollution, soil pollution, habitat destruction, biodiversity loss, and so forth. We face on a global scale a variety of environmental problems that in the end will have severe social impacts on our lives. It is also not just that climate change is being accelerated by human-induced development but that ecosystems, the climate and biodiversity are under threat from human development. We are coming closer to tipping points when small disturbances could change or undermine the fundamental ecological services that support all life on earth (Lenton *et al.*, 2008). David Schlosberg argues that there is a need to add a capability dimension to the environment in environmental justice; this would "enrich conceptions of environmental and climate justice by bringing recognition to the functioning of these systems, in addition to those who live within and depend on them" (Schlosberg, 2013: 44). Scholars such as John Drake and Reuben Keller (2004) and Mick Hillman (2006) have explored the importance of ecological integrity. There is a good argument that we need to have a change in our global understanding of development and that resource-intensive development threatens the environment through climate change and the biodiversity and ecological integrity of many habitats. We need to think of humans as dependent upon the environment as a functioning system rather than as a resource that should be exploited until it is gone. It mirrors the quote from Max Weber in the introduction that the capitalist economic world order will proceed until "the last ton of fossilized coal is burnt" (Weber, 1953: 181).

It is estimated that the populations of several terrestrial, aquatic and marine species have declined by more than 30 per cent since records were first kept in the early 1970s (WWF, 2010). There is a real threat that our ecosystems will not function and their integrity will be compromised to the extent that they will have difficulty in supporting life (Pimentel *et al.*, 2000; Earth Charter Initiative, 2010). Ecological integrity is important as it highlights the inherent potential, stability, capacity for self-repair and independent management of ecosystems (Karr, 1992). It is these features that enable ecosystems to provide, regulate and support all life (Millennium Ecosystem Assessment, 2005). This approach has created an attempt to include environmental concerns in an environmental justice movement that has often been perceived to be anthropocentric (Shrader-Frechette, 2002).

> When we interrupt, corrupt, or defile the potential functioning of ecological support systems, we do an injustice not only to human beings, but also to all of those non-humans that depend on the integrity of the system for their own functioning.
>
> (Schlosberg, 2013: 44)

This is an important dimension to add to environmental justice and low-carbon development. There is a need to look for new models of development if we want to solve some of the most fundamental environmental and social issues that we face on a global scale.

Low-carbon development and limits to consumption

We need to think about creating limits on the consumption of natural resources and to discuss how our levels of production and consumption threaten the capabilities of the environment. The consumption of natural resources is also a symbol for global social inequality. There is also an unsustainable disparity in consumption levels between richer and poorer countries. For example, the average Bangladeshi consumed 0.38 t CO_2 per capita in 2012, whereas the average American consumed 16.15 t CO_2 per capita, which is about 43 times higher (World Bank, 2012/13). There have been many attempts to discuss how humans could scale back their consumption and become more sustainable through taxation, regulations, investment, governance reform and pricing (Daly, 1977; Baer *et al.*, 2009; Stern, 2006; Sachs, 2008; Lohmann, 2009), but studies have not examined how these attempts might work in a nexus with ecological footprints on one side and social implications on the other. In addition to excessive carbon emissions, Johan Rockström and colleagues (2009) suggest that the planetary boundaries for biodiversity loss and nitrogen cycles may have been exceeded too. Regarding ocean acidification, the level of carbon emissions is also likely to play a major role (IPCC, 2013).

The debate about humanity's footprint on earth is also closely related to the concept of the earth's 'carrying capacity'. This is a concept that refers to how many people the earth can sustain in the long term. This relates to the earth's ability to provide resources for vital needs (such as food production), to withstand resource exploitation and change (such as climatic change) and to recover from shocks (such as pollution) (Ehrlich and Holdren, 1971; Daily and Ehrlich, 1996; Wackernagel *et al.*, 2002). Gemma Cranston and Geoff Hammond (2012) indicated that there is an imbalance between the inhabitants of the richer countries of the Global North, who live affluent lifestyles and over-shoot the earth's biocapacity with excessive carbon footprints, and those of the poorer countries of the Global South, who under-shoot the earth's biocapacity with modest carbon footprints. This raises questions of social justice in the climate change debate. This is partly being addressed through the discussions about 'loss and damage' that have dominated the international climate change negotiations since COP18 in Doha in late 2012.

The concept of the 'ecological footprint' is a statistical measure to evaluate the ecological impact of individual countries, communities or persons on the world's resources. It takes into account the amount of land, energy, water, forestry and other natural resources used and the emissions emitted, and it adds the trade of resources and access to modern technologies into the equation. The ecological footprint is expressed in terms of land requirements called 'global hectares' needed to sustain the living standards and consumption of an individual, a community or a country (Wackernagel and Rees, 1996). Global hectares

> are intended to reflect the areas of land and sea required to support production and consumption activities and assimilate waste materials. The estimates of global hectares are used for two purposes: 1. to compare the demand pressures imposed by individuals and communities, 2. to determine if the sum of demand pressures is greater than the available supply of the earth's productive and assimilative capacity.
>
> (Wichelns and Raina, 2011: 39)

The capacity of our planet to sustain humanity is calculated as 2.1 global hectares per capita, whereas the global average consumption was already 2.7 global hectares per capita back in the mid-1990s (Wackernagel and Rees, 1996; Holtzman, 1999; Amin, 2009). Jason Venetoulis and John Talberth suggested in 2008 that "humanity's footprint exceeds Earth's biocapacity by 23%" (Venetoulis and Talberth, 2008: 441). Recent research has refined and improved ecological footprint analysis (Venetoulis and Talberth, 2008; Wackernagel, 2009; Ruževi ius, 2011) and found that humanity's global footprint and its ecological over-shoot is in fact much greater than the standard approach by Wackernagel and Rees suggests (Venetoulis and Talberth, 2008). In addition, the rise of hundreds of millions of people belonging to the new middle class in emerging economies such as China and India has increased the overall average global footprint.

The challenge is therefore to address the disjunction between the twin aspirations of social policy to both protect the well-being of people worldwide and also to restrict excessive consumption among the richest countries and richest groups in low- and middle-income countries. This may lead to a redistribution and more equal use of natural resources on a global scale. Herman Daly suggests reducing growth and focusing more on having an "economy with constant population and constant stock of capital, maintained by a low rate of throughput that is within the regenerative and assimilative capacities of the ecosystem" (Daly, 2008: 3). He argues that there are three key policy tools that need to be in place to transform societies towards such a steady-state economy:

1 minimum and maximum limits on income and wealth;
2 improvements to the tax system;
3 restrictions on population growth.

In many ways, this brings us back to a main argument that low-carbon development can only be socially just if it means an honest and open discussion on the limits of growth and how resources need to be redistributed on a global scale to be more just and sustainable.

This book concludes that it might be time to decouple low-carbon growth from low-carbon development, as low-carbon development must strive for new models of economy and lifestyles. It is important that people are incentivised to "devote their further energies to noneconomic pursuits" so that confiscatory revenues would be rather small (Daly, 1977: 58) Scholars such as Larry Lohmann (2009), James Galbraith (2008) and Gus Speth (2008) argue not only for higher global carbon taxes and higher investment in low-carbon development, specifically low-carbon energy and low-carbon technology, but also suggest that there is a need to replace the market "by alternative democratic co-ordination and decision-making mechanisms" (Storm, 2009: 1026). The discourse around degrowth argues that endless economic growth is impossible to sustain. It questions the possibility of globally decoupling economic activity and emissions. In their famous book *The Limits to Growth*, Donella Meadows and colleagues argue that exponential growth in population and material output threatens the well-being of all and that it could lead to an uncontrolled global decline (Meadows *et al*., 1972). More recently, Tim Jackson (2009) and the Sustainable Development Commission (SDC) (2009) argue for "prosperity without growth" or prosperity within the ecological limits of our finite planet. A more progressive and democratic usage of resources will not come cheap and it will be dependent upon the willingness of people and governments to accept a model of lower consumption, lower growth, more equal distribution of resources and higher social equality that would result "in an improved welfare, a better quality of life and greater democratic control of production and (renewable) resources" (Storm, 2009: 1026).

This would need a complete reconfiguration of our understanding of welfare states, global social policy and the global market.

Note

1 The same shift in responsibility is not present among the devolved regions.

References

ACE (Association for the Conservation of Energy) (2014), *The Future of the Energy Company Obligation*, London: ACE, available at: www.ukace.org/wp-content/uploads/2014/04/ACE-Consultation-Response-2014-04-The-Future-of-the-Energy-Company-Obligation.pdf (accessed 26 March 2016).

Amin, S. (2009), "Capitalism and the ecological footprint", *Monthly Review*, 61(6), 19–22.

Antonelli, A. (1995), "Congress, not Clinton, supports the end of corporate welfare", Backgrounder Update No. 253, Washington, DC: Heritage Foundation.

Baer, P., Athanasiou, T., Kartha, S. and Kemp-Benedict, E. (2009), *The Greenhouse Development Rights Framework: The Right to Development in a Climate Constrained World*, revised second edition, Berlin: Heinrich Böll Stiftung, available at: www.ecoequity.org/docs/TheGDRsFramework.pdf (accessed 28 March 2016).

Barlett, D. L. and Steele, J. B. (1998), "Special report: exposing the folly of corporate welfare", *Time*, 152(19), 9 November.

Brenner, N. and Theodore. N. (2002), "Cities and the geographies of 'actually existing neoliberalism' ", in N. Brenner and N. Theodore (eds), *Spaces of Neoliberalism: Urban Restructuring in North America and Western Europe*, Oxford: Blackwell, pp. 349–79.

Cademartori, J. (2002), "Impacts of foreign investment on sustainable development in a Chilean mining region", *Natural Resources Forum*, 26(1), 27–44.

Carbon Trade Watch and FASE-ES (2013), *Like Oil and Water: Struggles Against the Brazilian Green Economy*, available at: www.carbontradewatch.org/multimedia/en/like-oil-and-water-struggles-against-the-brazilian-green-economy (accessed 8 July 2016).

Cranston, G. R. and Hammond, G. P. (2012), "Carbon footprints in a bipolar, climate-constrained world", *Ecological Indicators*, 16, 91–9.

Daily, G. C. and Ehrlich, P. R. (1996), "Socioeconomic equity, sustainability, and Earth's carrying capacity", *Ecological Applications*, 6(4), 991–1001.

Daly, H. (1977), *Steady-State Economics: The Political Economy of Bio-Physical Equilibrium and Moral Growth*, San Francisco, CA: W.H. Freeman & Co.

Daly, H. (2008), *A Steady-State Economy: A Failed Growth Economy and a Steady-State Economy Are Not the Same Thing; They Are the Very Different Alternatives We Face*, London: SDC.

Drake, J. M. and Keller, R. P. (2004), "Environmental justice alert: do developing nations bear the burden of risk for invasive species?", *Bioscience*, 54(8), 718–19.

Earth Charter Initiative (2010), *The Earth Charter*, San Jose: ECI, available at: http://earthcharter.org/discover/download-the-charter/ (accessed 23 March 2016).

Ehrlich, P. R. and Holdren, J. P. (1971), "Impact of population growth", *Science*, 171(3977), 1212–17.

Erlanger, S. and De Freytas-Tamura, K. (2015), "UN funding shortfalls and cuts in refugee aid fuel exodus to Europe", *New York Times*, 19 September, available at: www.nytimes.com/2015/09/20/world/un-funding-shortfalls-and-cuts-in-refugee-aid-fuel-exodus-to-europe.html?_r=0 (accessed 27 March 2016)

Freudenburg, W. R. (1992), "Vulnerable localities in a changing world economy", *Rural Sociology*, 57(3), 305–32.

Galbraith, J. K. (2008), *The Predator State: How Conservatives Abandoned the Free Market and Why Liberals Should Too*, New York: Free Press.

Gilbertson, T. and Reyes, O. (2009), *Carbon Trading: How It Works and Why It Fails*, Uppsala: Dag Hammarskjöld Foundation.

The Green Age (2015), "Why did the green deal fail?", available at: www.thegreen-age.co.uk/why-did-the-green-deal-fail/ (accessed 26 March 2016).

Grupo Plantar (n.d.), website, available at: www.grupoplantar.com.br/ (accessed 23 March 2016).

Hamilton, I., Shipworth, D., Summerfield, A., Steadman, P., Oreszczyn, T. and Lowe, R. (2014), "Uptake of energy efficiency interventions in English dwellings", *Building Research and Information*, 42(3), 255–75.

Hayashi, D. and Michaelowa, A. (2013), "Standardization of baseline and additionality determination under the CDM", *Climate Policy*, 13(2), 191–209.

Hillman, M. (2006), "Situated justice in environmental decision-making: lessons from river management in southeastern Australia", *Geoforum*, 37(5), 695–707.

Hills, J. (2012), *Getting the Measure of Fuel Poverty: Final Report of the Fuel Poverty Review*, London: Centre for Analysis of Social Exclusion, available at: http://sticerd.lse.ac.uk/dps/case/cr/CASEreport72.pdf (accessed 26 March 2016).

Holtzman, D. (1999), "Ecological footprints", *Dollars & Sense*, July/August, 224, 42.

International Rivers (2015), *Benchmarking the Policies and Practices of International Hydropower Companies: Stage 1 – Environmental and Social Policies and Practices of Chinese Overseas Hydropower Companies*, Berkeley, CA: International Rivers, available at: www.internationalrivers.org/files/attached-files/benchmarking_report_english_part_a.pdf (accessed 20 March 2016).

IPCC (Intergovernmental Panel on Climate Change) (2013), *Working Group I Contribution to the Fifth Assessment Report: The Physical Science Basis*, Geneva: IPCC, available at: www.climatechange2013.org/images/uploads/WGIAR5_WGI-12Doc2b_FinalDraft_All.pdf (accessed 13 March 2016).

IWR (German Institute of the Renewable Energy Industry) (2014), "Renewable energies are subsidised – the state doesn't pay a cent" (in German), Münster: IWR, available at: www.iwr-institut.de/de/presse/presseinfos-energiewende/erneuerbare-energien-werden-subventioniert-staat-zahlt-keinen-cent#wie-die-EEG-Umlage-funktioniert (accessed 22 March 2016).

Jackson, T. (2009), *Prosperity Without Growth*, London: Earthscan.

JWW (Jewish World Watch) (2015), "Solar Cooker Project: food crisis in the Darfuri refugee camps", Encino: JWW, available at: www.jww.org/projects/ontheground/solar-cooker-project-redirect (accessed 8 July 2016).

Karr, J. R. (1992), "Ecological integrity: protecting earth's life support systems", in R. Costanza, B. Norton and B. Haskell (eds), *Ecosystem Health: New Goals for Environmental Management*, Washington, DC: Island Press, pp. 223–38.

Lindsey, B. (1992), "Sematech: the wrong solution", *Journal of Commerce*, 16 January.

Lenton, T. M., Held, H., Kriegler, E., Hall, J. W., Lucht, W., Rahmstorf, S. and Schellnhuber, H. J. (2008), "Tipping elements in the Earth's climate system", *Proceedings of the National Academy of Sciences of the United States of America*, 105(6), 1786–93.

Lohmann, L. (2006), "Carbon trading: a critical conversation on climate change, privatisation and power", *Development Dialogue*, 48, 1–362.

Lohmann, L. (2009), "Climate as investment", *Development and Change*, 40(6), 1063–83.

Mallaburn, P. and Eyre, N. (2014), "Lessons from energy efficiency policy and programmes in the UK from 1973 to 2013", *Energy Efficiency*, 7(1), 23–41.

Meadows, D. H., Meadows, D. L., Randers, J. and Behrens III, W. W. (1972), *The Limits to Growth: A Report to the Club of Rome*, New York: Universe Books.

Michaelowa, A. and Purohit, P. (2007), "Additionality determination of Indian CDM projects", Discussion Paper CDM-1, Climate Strategies, available at: www.internationalrivers.org/files/attached-files/additionality-cdm-india-cs-version9-07.pdf (accessed 8 July 2016).

Millennium Ecosystem Assessment (2005), *Ecosystems and Human Well-Being: Synthesis*, Washington, DC: Island Press, available at: www.millenniumassessment.org/documents/document.356.aspx.pdf (accessed 23 March 2016).

Moore, S. and Stansel, D. (1995), "Ending corporate welfare as we know it", Cato Policy Analysis No. 225, Washington, DC: Cato Institute.

Pearce, F. (2002), "Tree farms won't halt climate change", *New Scientist*, available at: www.newscientist.com/article/dn2958-tree-farms-wont-halt-climate-change/ (accessed 8 July 2016).

Pearse, R. and Böhm, S. (2014), "Ten reasons why carbon markets will not bring about radical emissions reduction", *Carbon Management*, 5(4), 325–37.

Pimentel, D., Westra, L. and Noss, R. (2000), *Ecological Integrity: Integrating Environment, Conservation and Health*, Washington, DC: Island Press.

Richards, J. (2012), "The Green Deal", *Mortgage Finance Gazette*, 6 December, available at: www.mortgagefinancegazette.com/legal/the-green-deal/ (accessed 26 March 2016).

Rockström, J., Steffen, W. L., Noone, K. *et al.* (2009), "Planetary boundaries: exploring the safe operating space for humanity", *Ecology and Society*, 14(2), 32.

Ruževičius, J. (2011), "Ecological footprint: evaluation methodology and international benchmarking", *Business & Law*, 6(1), 11–30.

Sachs, J. (2008), "Technological keys to climate protection", *Scientific American*, 298(4), 40.

Schatz, T. A. (1997), "Prepared testimony of Thomas A. Schatz, president, Citizens Against Government Waste, before the Senate Committee on Governmental Affairs", Washington, DC: US Senate.

Schlosberg, D. (2013), "Theorising environmental justice: the expanding sphere of a discourse", *Environmental Politics*, 22(1), 37–55.

Shrader-Frechette, K. (2002), *Environmental Justice: Creating Equality, Reclaiming Democracy*, New York: Oxford University Press.

Speth, J. G. (2008), *The Bridge at the Edge of the World: Capitalism, the Environment and Crossing from Crisis to Sustainability*, New Haven, CT and London: Yale University Press.

Stern, N. (2006), *Stern Review: The Economics of Climate Change*, London: HM Treasury, available at: http://webarchive.nationalarchives.gov.uk/20100407172811/www.hm-treasury.gov.uk/stern_review_report.htm (accessed 15 March 2016).

Storm, S. (2009), "Capitalism and climate change: can the invisible hand adjust the natural thermostat?", *Development and Change*, 40(6), 1011–38.

SDC (Sustainable Development Commission) (2009), *Prosperity Without Growth*, London: SDC.

UNFCCC (United Nations Framework Convention on Climate Change) (2005), 9th Plenary Meeting, 3/CMP.1, Annex.

Urban, F. and Lind, J. (2011), "Low carbon energy and conflict: a new agenda", *Boiling Point*, 59, available at: www.hedon.info/BP59_Low+carbon+energy+and+conflict (accessed 27 March 2016).

Urban, F., Nordensvärd, J., Khatri, D. and Wang, Y. (2013), "An analysis of China's investment in the hydropower sector in the Greater Mekong Subregion", *Environment, Development and Sustainability*, 15(2), 301–24.

Urban, F., Nordensvärd, J., Siciliano, G. and Li, B. (2015), "Chinese overseas hydropower dams and social sustainability: the Bui Dam in Ghana and the Kamchay Dam in Cambodia", *Asia & the Pacific Policy Studies*, 2(3), 573–89.

Vaughan, A. (2013), "Green Deal's upfront fees 'put people off upgrading homes'", *Guardian*, 6 January, available at: www.theguardian.com/environment/2013/jan/06/green-deal-upfront-fees-upgrading (accessed 26 March 2015).

Venetoulis, J. and Talberth, J. (2008), "Refining the ecological footprint", *Environment, Development and Sustainability*, 10(4), 441–69.

Wackernagel, M. (2009), "Introduction: methodological advancements in footprint analysis", *Ecological Economics*, 68(7), 1925–7.

Wackernagel, M. and Rees, W. E. (1996), *Our Ecological Footprint: Reducing Human Impact on the Earth*, Gabriola Island: New Society.

Wackernagel, M., Schulz, N. B., Deumling, D., Linares, A. C., Jenkins, M., Kapos, V., Monfreda, C., Loh, J., Myers, N., Norgaard, R. and Randers, J. (2002), "Tracking the ecological overshoot of the human economy", *Proceedings of the National Academy of Sciences of the United States of America*, 99(14), 9266–71.

Weber, M. (1953), *The Protestant Ethic and the Spirit of Capitalism*, New York: Scribner.

Wichelns, D. and Raina, A. (2011), "Would water footprints enhance water policy in India?", *The Water Digest*, 2011, 32–44, available at: http://lkyspp.nus.edu.sg/iwp/wp-content/uploads/sites/3/2013/10/water_digest_aditi_dw_india_131016.pdf (accessed 9 June 2016).

World Bank (2012/13), "World development indicators", available at: http://data.worldbank.org/data-catalog/world-development-indicators (accessed 28 March 2016).

WWF (World Wildlife Fund) (2010), *Living Planet Report*, Gland: WWF, available at: http://d2ouvy59p0dg6k.cloudfront.net/downloads/wwf_lpr2010_lr_en.pdf (accessed 9 June 2016).

Index

Page numbers in *italics* denote tables, those in **bold** denote figures.

Helping you to choose the right eBooks for your Library

Add Routledge titles to your library's digital collection today. Taylor and Francis ebooks contains over 50,000 titles in the Humanities, Social Sciences, Behavioural Sciences, Built Environment and Law.

Choose from a range of subject packages or create your own!

Benefits for you

- Free MARC records
- COUNTER-compliant usage statistics
- Flexible purchase and pricing options
- All titles DRM-free.

Free Trials Available
We offer free trials to qualifying academic, corporate and government customers.

Benefits for your user

- Off-site, anytime access via Athens or referring URL
- Print or copy pages or chapters
- Full content search
- Bookmark, highlight and annotate text
- Access to thousands of pages of quality research at the click of a button.

eCollections – Choose from over 30 subject eCollections, including:

- Archaeology
- Architecture
- Asian Studies
- Business & Management
- Classical Studies
- Construction
- Creative & Media Arts
- Criminology & Criminal Justice
- Economics
- Education
- Energy
- Engineering
- English Language & Linguistics
- Environment & Sustainability
- Geography
- Health Studies
- History
- Language Learning
- Law
- Literature
- Media & Communication
- Middle East Studies
- Music
- Philosophy
- Planning
- Politics
- Psychology & Mental Health
- Religion
- Security
- Social Work
- Sociology
- Sport
- Theatre & Performance
- Tourism, Hospitality & Events

For more information, pricing enquiries or to order a free trial, please contact your local sales team: www.tandfebooks.com/page/sales

The home of Routledge books

www.tandfebooks.com